AF302491

Reactivity and Structure Concepts in Organic Chemistry

Volume 8

Editors:

Klaus Hafner Jean-Marie Lehn
Charles W. Rees P. von Ragué Schleyer
Barry M. Trost Rudolf Zahradník

Christian Birr

Aspects of the Merrifield Peptide Synthesis

With 62 Figures and 6 Tables

Springer-Verlag
Berlin Heidelberg New York 1978

Priv.-Doz. Dr. Christian Birr

Max-Planck-Institut für medizinische Forschung
Abteilung Naturstoff-Chemie
Jahnstr. 29, D-6900 Heidelberg 1

ISBN-13: 978-3-642-67006-0   e-ISBN-13: 978-3-642-67004-6
DOI: 10.1007/978-3-642-67004-6

Library of Congress Cataloging in Publication Data. Birr, Christian, 1937– Aspects of the Merrifield peptides synthesis. (Reactivity and structure ; v. 8) Bibliography: p. Includes index. 1. Peptide synthesis. I. Title. II. Series. QD431.B57   547'.756   78-16659

© by Springer-Verlag Berlin Heidelberg 1978
Softcover reprint of the hardcover 1st edition 1987

Typesetting: Elsner & Behrens, Oftersheim

2152/3140-5 4 3 2 1 0

Preface

This book was written in the context of the daily confrontation with problems in the utilization of polymeric supports for the synthesis of peptides. Therefore, views and experiences which usually are not mentioned in scientific journals are collected in these pages. The author has deliberately discussed in detail the possible influence of the polymer phase on the varying reaction conditions in the Merrifield synthesis; this aspect is neglected in most publications dealing with peptide synthesis. However, in view of the growing body of information on the chemistry of polymer-supported peptide syntheses, the international readership should regard the author's arguments as open to discussion.

I am very much indebted to all of my colleagues with whom I have had the opportunity to cooperate in studying the potential of the Merrifield synthesis. Above all I like to express my gratitude to my teacher, Professor Dr. Theodor Wieland, Heidelberg, for his boundless encouragement and support in my efforts in the field of peptide synthesis, particularly in its polymer phase bound version. Last but not at least I wish to thank Miss Hildegard Leyden. With infinite patience and great accuracy she typed the manuscript in addition to her daily duties.

The work which influenced this book was performed under the auspices of the Max-Planck-Gesellschaft, of which the financial support is gratefully acknowledged.

<table>
<tr><td>Heidelberg, June 1978</td><td align="right">Christian Birr</td></tr>
</table>

Contents

Contents

1 Introduction

In most areas of science, knowledge seems to increase not in a continuous process but in a way of certain quanta jumps, from a level of insight to the next better. Surely this is the case in the development of peptide chemistry.

From my point of view, the evolution of the methodology of peptide synthesis may be characterized by four distinct periods. Although syntheses of amino acids and their reactivity were well known before, the finding of E. Fischer [1] in 1901–1903 of how to transform amino acids to render a peptide synthesis possible marked the breakthrough to protein research.

But not until 1932, when M. Bergmann and L. Zervas [2] published the first protecting group for temporary blocking of the N-terminal amino function during peptide bond formation, was a second level in the knowledge of methods in peptide synthesis reached. The use of the carbobenzoxy group considerably simplified the elongation of peptide chains. Consequently the very versatile principle of the urethane type of protection of amino functions was often varied and adapted to very different reaction conditions during the next two decades.

Doubtless, a new field in the development of peptide synthesis was opened when Th. Wieland (1950–51) [3] published the methods of activation of amino acid carboxylic functions both by formation of mixed anhydrides with ethylchlorocarbonate and by active esters with thiophenol [4]. At the same time also, R. Boissonnas [5] and J. R. Vaughan [6] independently described the mixed anhydride method. A most fruitful period of peptide synthesis began, which culminated (for example) in the synthesis of peptide hormones like oxytocin by duVigneaud (1953) [7], ACTH by R. Schwyzer (1963) [8], and insulin by the group of H. Zahn in 1963 [9]. Besides these examples, hundreds of important syntheses of natural occurring peptides were published in that decade, but it would go beyond the scope of this introduction to note even some of them. Although further protecting groups were introduced, and the methods of peptide coupling became more and more subtilized, recognition increased that one has to search for a new principle of peptide synthesis, to reach the goal of building up peptide chains of a hundred or more amino acid residues, by copying nature's principles, to synthesize proteins.

In the beginning sixties, when R. B. Merrifield for the first time described his new idea of solid phase peptide synthesis (1962) [10], the cellular mechanism of ribosomal protein biosynthesis was already known, and was no longer a subject of hypothetic discussions. The ribosomal area of the biological stepwise condensation of amino acids might be described by the term "gel phase", where reactions proceed between solution and solid phase in a region to be paraphrased by hydrophobic-hydrophilic interplay. It seems plausible to me that Merrifield's idea of solid phase peptide synthesis was influenced by the knowledge of this biological principle of surface reactions on a gelatinous phase. Later on we shall see that from the chemical point of view, the Merrifield peptide synthesis differs drastically from the very rapid reactions of the protein biosynthesis.

Since 1962, Merrifield's new idea induced unanticipated activity in peptide chemistry around the world. The synthesis of artificial enzymes and hormones seemed to become possible, and hundreds of organic chemists and biochemists, as well as enzymologists and physicians, devoted themselves to the new method. Unfortunately the potential of the solid phase peptide synthesis was, from the very beginning, erroneously estimated by the international scientific public, both in a positive and negative sense. By emotional arguments more than by objective ones, the peptide chemists until recently were split into two parties, of which one trusted and the other condemned the solid phase peptide synthesis.

The enthusiastic belief in the new method promising rapid success in protein synthesis led to some publications which were in want of the necessary critical care. On the other hand, the methodology of the solid phase synthesis was elaborated and improved by a great number of peptide research groups. Hundreds of excellent investigations, not only concerning chemical details of the new technique, but especially peptide syntheses on solid phase, have been published till today. This progress in the solid phase method of Merrifield is reviewed in several articles and most detailed and comprehensive books [11—35]. In the book submitted here, I therefore shall confine myself to evaluating the chemical aspects of the Merrifield method from the standpoint of present practical use. Several developments of methodical details will be balanced critically against each other, without the intention of reviewing the literature completely or discussing the fundamentals of peptide synthesis from the very beginning.

2 The Principle

Contrary to the principles of the ribosomal protein biosynthesis, where the carboxylic functions of amino acids are reactively bound to a polymer — the transfer ribonucleic acid (t-RNA) — which itself is orientated specifically on the ribosomal messenger ribonucleic acid (m-RNA) (Fig. 1; for details see [36]), the basic idea of the Merrifield synthesis depends upon a nonreactive covalent fixation of the C-terminal amino acid of the target peptide on a solid support (Fig. 2). On this insoluble but swollen polymer, the pep-

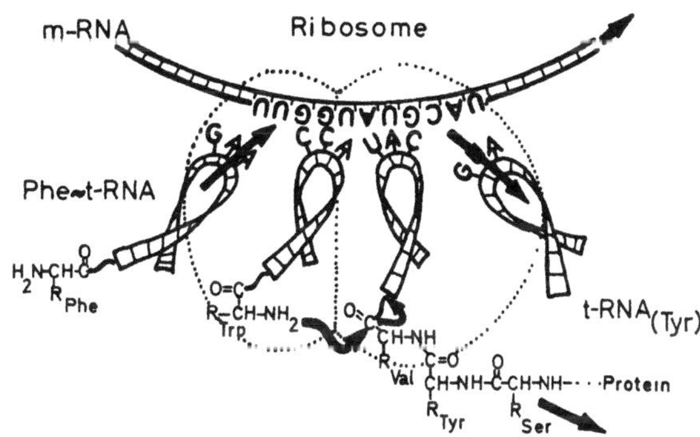

Fig. 1. The ribosomal area of the biological stepwise condensation of amino acids in the protein biosynthesis

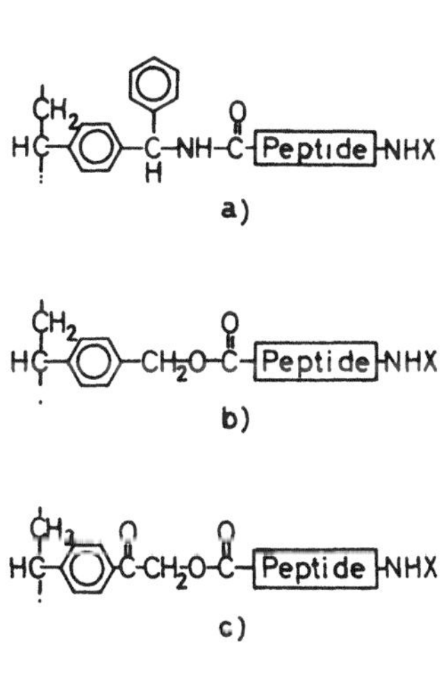

Fig. 2. Three examples of the solid anchoring of amino acids and peptides in polymer phase, nonreactive during chain elongation in the Merrifield peptide synthesis [type a) benzhydryl amine, b) benzyl ester and c) phenacetyl ester]

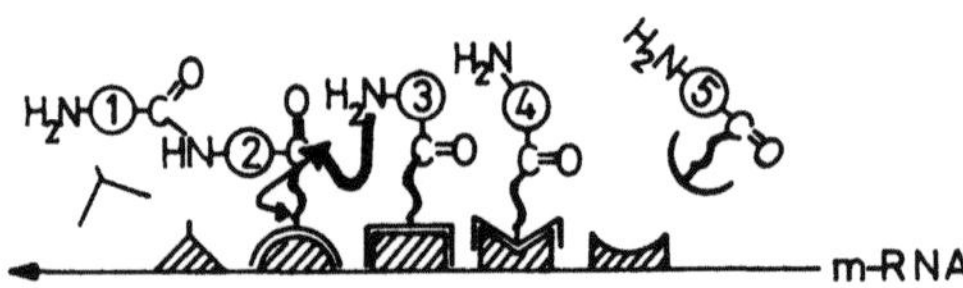

Fig. 3. The aminolytic chain elongation in the protein biosynthesis

tide to be synthesized is assembled in a stepwise manner by N-terminal elongation of the chain with N-protected and C-activated amino acids, which are the dissolved carboxyl components of the solid phase peptide synthesis.

In contradiction to this, the protein biosynthesis proceeds via aminolysis of activated amino acids reactively bound to t-RNA, where the growing protein is liberated from the polymer RNA support (Fig. 3). This type of biosynthesis was imitated in the way of a drastically simplified model reaction, namely the peptide syntheses with solid phase bound activated amino acid esters [37, 38], which here are not the matter for discussion (Fig. 4).

On the first view, the brilliant idea of Merrifield to synthesize peptides onto an insoluble polymer support seems to be as ingenious as it is simple (Fig. 5). The first amino acid, insolubilized by linkage to a suitable polymer support, is deprotected at its amino function and reacted with the next N-protected amino acid to form the first peptide bond. Necessary reagents and excessive N-protected amino acid derivatives are thoroughly removed by repeated washings of the polymer-bound peptide with appropriate solvents. This sequence of operations is repeated during each peptide elongation stage. At the end

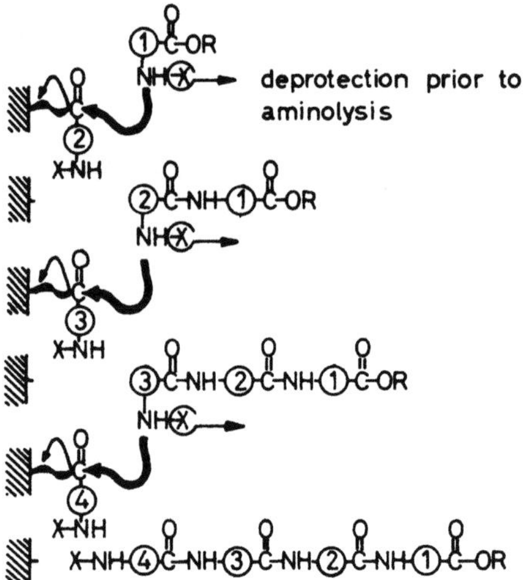

Fig. 4. The stepwise peptide synthesis in gel phase with polymer-supported active esters of amino acids, a model of the protein biosynthesis

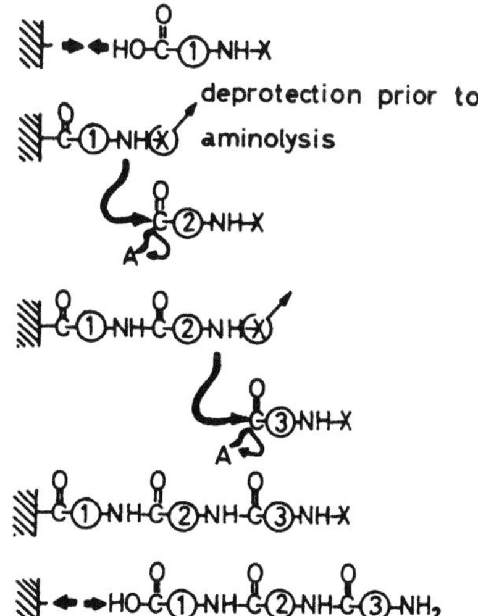

Fig. 5. The principal scheme of the Merrifield solid phase peptide synthesis

of the synthesis on solid phase, the link of the peptide to its polymer support is cleaved, the dissolved peptide is separated from the insoluble resin and has to be purified. It is obvious that this procedure extremely simplifies the manipulations of classical peptide chemistry, because it utilizes the same classical reactions for deprotection, deprotonation, and activation of amino acids, as well as for cleavage of the peptide from the support. All the time-consuming and wasteful operations in conventional peptide synthesis, to derivatize intermediates and to work-up and purify these products by extractions and crystallization, as well as to analyze and characterize each intermediate peptide combined with often repeated transfers of the synthetic materials through several glass-vessels, are eliminated in the Merrifield synthesis, or are substituted for simple intense washings and filtration of the polymer bound peptide in a single reaction vessel.

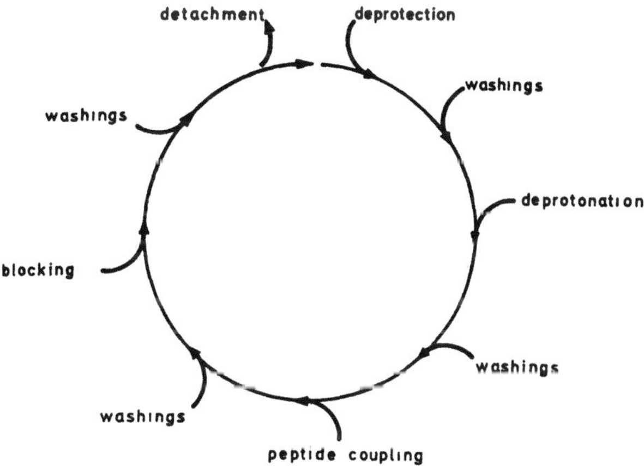

Fig. 6. The cyclic sequence of operations in the Merrifield peptide synthesis

The sequence of reactions, washings, and filtrations necessary for one peptide bond formation is called "cycle" (Fig. 6) and consists of several "steps", which are, for example,

deprotection,
deprotonation,
peptide bond formation,
washings, and
filtration.

Each step from the chemical and practical view has its own problems and therefore will be discussed in detail later on in separate sections. The succession of "cycles" during a solid phase synthesis characterizes the various "stages" of the peptide elongation. Since all operations in the Merrifield method are often repeated, the procedure is predestined for automatic peptide synthesis. Consequently the first apparatus for this purpose had already been constructed in the incipience of the methodical development by R. B. Merrifield and J. M. Stuart [39] (see p. 72).

2.1 Aspects of the Strategy

Merrifield's original idea was based on the general scheme of stepwise condensation of N-protected amino acids to the first one, which is linked with its carboxyl function by an ester bond to the insoluble polymer support. This way of solid phase peptide synthesis resulted from the well-known risk of racemization during activation of peptidic carboxyl components, which is minimized in activated amino acid derivatives, N-acylated by urethane-type protecting groups [40] (Fig. 7). Depending on the chosen method, the C-terminal activation of N-protected peptides tends to racemize a certain amount of the material because of the possible formation of an oxazolinone intermediate [41] (Fig. 8).

For that reason, the applicability of the other strategy introduced by R. L. Letsinger and M. J. Kornet in 1963 [42] is in contrast to the Merrifield procedure limited. In the second method the stepwise peptide synthesis starts from the opposite direction, beginning from the N-terminal amino acid bound by its amino function to an insoluble support. The peptide is elongated by activations of the solid phase carboxyl component on each

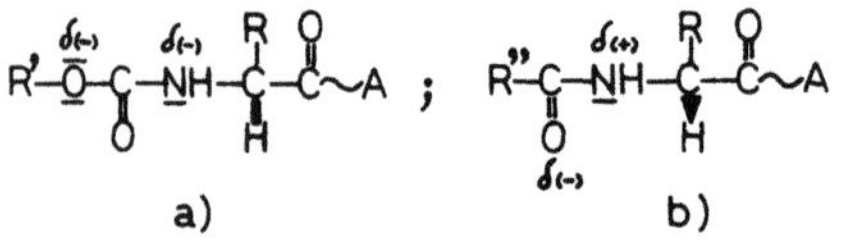

Fig. 7. The urethane type of N-protecting groups (a) shields carboxyl activated amino acids against racemization, in contrast to common N-acyl residues (b)

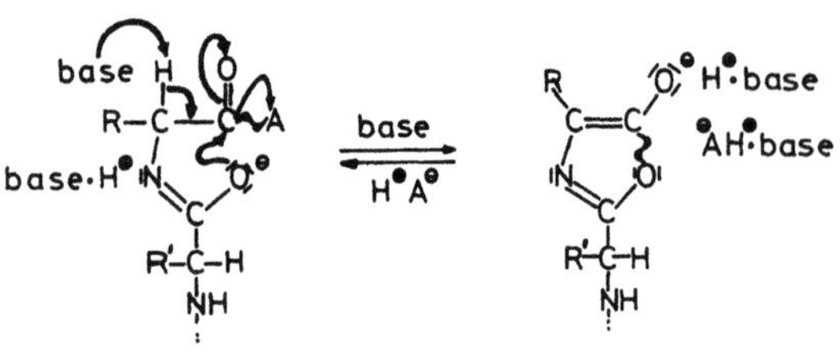

Fig. 8. The oxazolin-5-one mechanism of the C-terminal peptide racemization

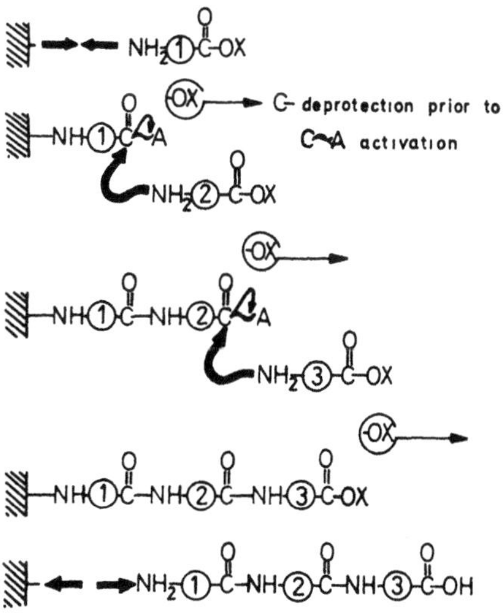

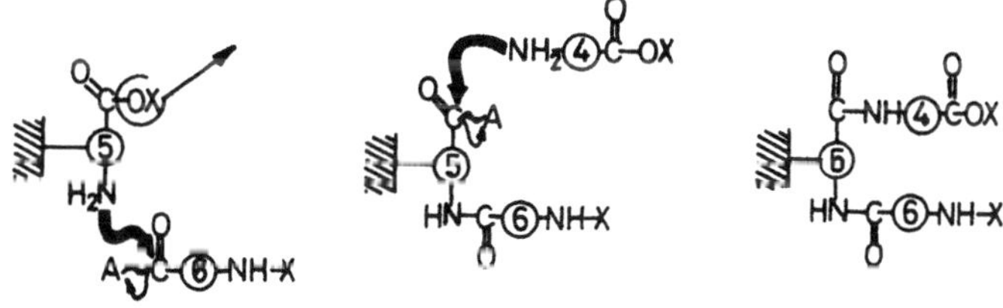

Fig. 9. The schematic peptide synthesis on polymer by C-terminal chain elongation according to Letsinger and Kornet

stage of the stepwise synthesis (Fig. 9). To what extent the utilization of this method might be improved by racemization suppressing additives [43] like 1-hydroxybenzotriazole and others, which recently have been introduced by R. Geiger and W. König [44], is not the matter for discussion here. Rather, I want to relate to another approach of the stepwise peptide synthesis on solid phase, which combines the potential of Merrifield's and Letsinger's strategies. Since almost fifty percent of the naturally occuring amino acids are trifunctional ones like glutamic acid, asparagine, tyrosine, serine, cysteine, histidine, lysine and others, a starting amino acid for the purpose of solid phase peptide synthesis can be linked covalently to the polymer support by its third function [45—48]. This aspect not only includes the possibility of peptide synthesis in two directions (Fig. 10), but also is advantageous in peptide cyclization on solid phase (Fig. 11). To consider the polymer support a special reagent to protect, for example, the sulfhydryl function of cysteine or cysteine containing peptides [49], this view opens peculiar chances to synthesize peptides for subsequent disulfide cyclization. J. D. Glass, R. Walter, and J. L. Schwartz [50] demonstrated a further variation of binding of trifunctional amino acids like histidine by using a dinitrofluorophenyl-modified polymer support (Fig. 12) for bidirectional solid phase synthesis. The method promises mild conditions of peptide cleavage from the polymer and therefore will be discussed in a separate chapter. From the strategic point of view, it is hard to evaluate the advantage of this bridging method because of the early stage of

Fig. 10. The bidirectional peptide synthesis in polymer phase

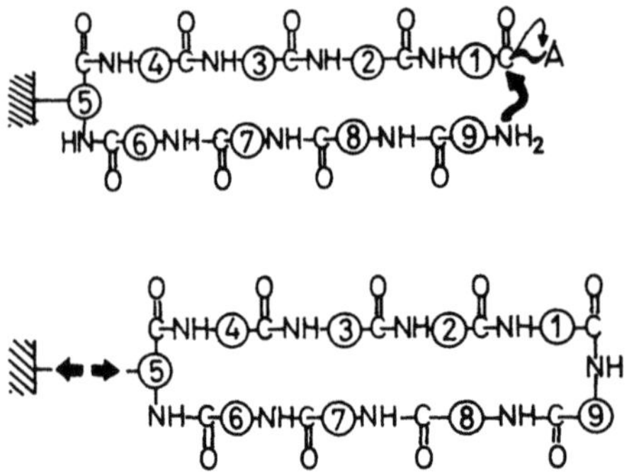

Fig. 11. The peptide cyclization in polymer phase after bidirectional synthesis of the linear precursor

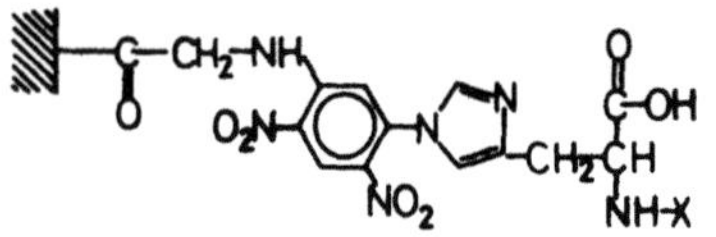

Fig. 12. The 2,4-dinitrophenylene bridge links histidine to the modified support for bidirectional synthesis

experiences. Obviously it is still problematic to transform in high yield the carboxyl functions of peptides which are bound to a polymer support into activated esters or other intermediates of sufficient reactivity, without aggravating the racemization tendency.

Although this argument is concerned even more with the activation of dissolved peptides, the original strategy of Merrifield involving stepwise amino acid condensation is being modified occasionally in terms of stepwise peptide fragment condensation on solid phase, which differs from the conventional technique in dissolved systems, because of the exclusive N-terminal peptide elongation on polymer support (Fig. 13). As discussed in the

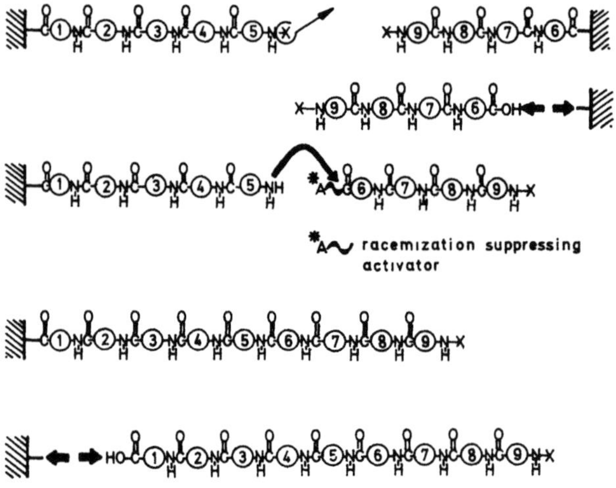

Fig. 13. Scheme of the N-terminal peptide elongation in polymer phase by stepwise fragment condensation

following section, this modification has its extraordinary importance in questions of possible syntheses of pure proteins – peptides consisting of more than 100 amino acid residues – on solid phase, because of the enlarged molecular weight of the peptidic building blocks in stepwise fragment condensation, compared to amino acid units in the original strategy of Merrifield.

2.2 The Statistical Point of View on Solid Phase Reactions

Behind the elegance and the ingenious simplicity of Merrifield's idea of peptide synthesis a serious problem is hiding, which is inherent in the principle itself, namely the demand for completion of all chemical reactions on solid phase [51].

In the early stage of application of the Merrifield method to polypeptide syntheses, either this supreme requirement was not perceived or its consequences were disregarded. This sin of omission, on the one hand, fundamentally discredited the method temporarily with groups of international peptide chemists, as indicated in the Introduction. But on the other hand, from its very first enunciation, this demand signified a strong provocation to numerous organic chemists engaged in peptide syntheses to overcome several shortcomings of the Merrifield method. These shortcomings are derived from the difficulty of realizing complete reactions on a polymer support and therefore will be the matter for discussion in diverse sections of this book.

The interpretation of the problem depends upon some statistical aspects of chemical transformations on insoluble polymers. Contrary to the homogeneous situation in real dis-

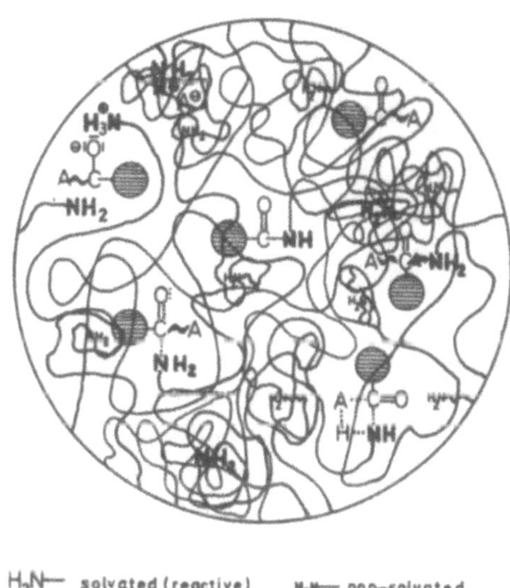

Fig. 14. The statistical distribution of functional transformations during peptide synthesis in heterogeneous gel phase

solved phases, the bimolecular reaction conditions in a Merrifield support vary by different microenvironments of the reactive functions in the inhomogeneous polymer matrix (for further discussion see Sect. 3.1, page 16). This fact results not only in slower reaction rates by phase separation in general, but in a statistical distribution of the degree of all transformations under distinct conditions for each step of a solid phase peptide synthesis (Fig. 14). For this reason it is very difficult to reach the goal of complete transformations on an insoluble support, and even more difficult to analyze the reaction since the same uncertainty of reaction conditions seriously concerns the direct analytical control of 100% conversions on solid phase by additional chemical reactions.

During the ninth European Peptide Symposium 1968, in Orsay, France, the reasonable view was published [52] that even very small deviations — less than 1% — of reaction rates from completion (a very optimistic assumption as we already know from reactions in real dissolved systems) have to generate a statistical distribution of by-products accompanying the target peptide. To get an impression of these inhomogeneities which depend on factors such as incomplete deprotection, deprotonations, and peptide bond formations on solid phase, several groups of peptide chemists tried to calculate these effects roughly as a function of constant yield of peptide elongations in each cycle of a Merrifield synthesis. We compared our results [53] with those of E. Bayer's group — which were published later [54] — and found equivalent expressions on the basis of the binominial proposition.

The yield $y(\%)$ of each cycle c of a synthesis of n stages may deviate from complete conversions (100%) in a constant failure of $f(\%)$,

$$y(\%) = 100 - f(\%), \quad c = y + f = 1$$

The product composition after n cycles c of the synthesis is then described by the binominial expression,

$$c^n = (y + f)^n$$

which is resolved to

$$c^n = y^n + nfy^{(n-1)} + \frac{1}{2} n(n-1)^2 f^2 y^{(n-2)} + \frac{1}{6} n(n-1)(n-2)^2 f^3 y^{(n-3)} + \ldots$$

$$+ \; nyf^{(n-1)} + f^n$$

wherein the terms of c^n characterize the additive statistical distribution of defective peptide sequences in molefractions accompanied by the yield of the target peptide.

For practical reasons we inspect only the first three members of the equation, since they present a quick impression of the theoretical course of a stepwise synthesis, and should stimulate each interested reader to examine synthetic problems of his own choice.

The term y^n indicates the yield of the target peptide, which has a chain length of $(n+1)$ amino acid residues. The member $nfy^{(n-1)}$ represents the yield of all defective sequences of n residues length, whereas $\frac{1}{2} n(n-1)^2 f^2 y^{(n-2)}$ equals the yield of all false sequences with $(n-1)$ amino acid residues, and so forth.

Table 1. Theoretical end product composition of an incremental peptide synthesis of 100 stages with a constant failure of 1% per peptide elongation

Chain lengths	Sequence Variations V	Yield (%)
101[a]	0[a]	36
100[b]	100	36[c]
99[b]	5,000[c]	18[c]
98[b]	160,000[c]	6[c]
97[b]	4,000,000[c]	1.5[c]
96[b]	80,000,000[c]	0.3[c]

[a] Target peptide.   [b] By-products.   [c] Rounded off.

Expressed in real numbers a Merrifield synthesis involving, for example, $n = 100$ stages with a constant failure $f = 1\%$ in each cycle would yield the target peptide containing 101 residues in 36% (since $y^n = 0.99^{100} = 0.36$), mixed up with by-product sequences of 100 residues length in a molefraction of likewise about 36% (namely $nfy^{(n-1)} = 100 \cdot 0.01 \cdot 0.99^{99} \sim 0.36$), and with by-products combining all possible sequences with a chain length of 99 amino acid residues in 18%.

Since the quantity of all possible sequence variations V in each additive member of the binominial equation, which symbolizes the molefraction of the by-product group X (all sequences X residues shorter than the target peptide. $X = $ exp. of $f = 0, 1, 2 ...$) in a synthesis of n stages, is indicated by $V = n!/X! (n - X)!$, the theoretical composition of the above-mentioned stepwise synthesis is roughly determined by the figures of Table 1.

The fundamentality of the demand for completion of all reactions is saliently elucidated within the background of these simple estimations on the theoretical result of a solid phase synthesis. It is obviously impossible to separate about 84,165 million extremely similar peptidic by-products — whose molecular weights do not differ by more than five amino acid residues of weight — from the main sequence by any presently available method of purification.

From more accurate calculations of J. M. A. Baas et al. [55] of theoretical end product compositions of natural occurring peptides (Table 2) — which in the meantime nearly all were taken as goals for solid phase syntheses — one has to appreciate that Merrifield syntheses of sequences larger than 15—25 amino acid residues, without guaranteed yield of at least 99.5% in each stage of the procedure must result in peptide end products of indefinite chemical composition and of uncertain scientific value, even though one confines the judgement on biological activity of the material in question: The target peptide, e.g., a synthetic enzyme, a hormone or even parts of these, may contain millions of closely related defective sequences of similar biological activities. Without knowledge of any distinct sequence, the additive activities of these by-products create the impression of the success of the synthesis. The target peptide, however, may scarcely range in the material.

In spite of these murky aspects, which concern the adventures in the field of syntheses of polypeptides and proteins on solid phase by incremental chain elongation, the method of Merrifield is invaluable in the quick and convenient preparation of smaller peptides

Table 2. The postulated end product composition related to deviations from completed stepwise reactions in the Merrifield synthesis of target sequences from natural origin, according to Baas et al. [55]

Relative statistical distribution of peptides. 8 coupling steps; e.g., oxytocin

Number of amino acids	Coupling yield (%)	Molefraction in % for 8 coupling steps					
		90	95	98	99	99.5	99.9
9[a]		43.0	66.3	85.1	92.3	96.1	99.2
8[b]		38.3	27.9	13.9	7.5	3.9	0.8
7[c]		14.9	5.1	1.0	0.3	0.1	
6[c]		3.3	0.5				
5[c]		0.5					

[a] One nonapeptide only.
[b] Total molefraction of *all possible* (octa) peptide combinations is meant.   [c] Idem.

91 coupling steps; e.g., neocarzinostatin

Number of amino acids	Coupling yield (%)	Molefraction in % for 91 steps			
		98	99	99.5	99.9
92		15.9	40.1	63.4	91.3
91		29.5	36.8	29.0	8.3
90		27.1	16.7	6.6	0.4
89		16.4	5.0	1.0	
88		7.4	1.1	0.1	
87		2.6	0.2		
86		0.8			
85		0.2			

123 coupling steps; e.g., ribonuclease

Number of amino acids	Coupling yield (%)	Molefraction in % for 123 steps			
		98	99	99.5	99.9
124		8.3	29.0	54.0	88.4
123		20.9	36.1	33.4	10.9
122		26.0	22.2	10.2	0.7
121		21.4	9.1	2.1	
120		13.1	2.7	0.3	
119		6.4	0.7		
118		2.6	0.1		
117		0.9			
116		0.3			
115		0.1			

Table 2 (continued)

54 coupling steps; e.g., apoferredoxin

Number of amino acids	Coupling yield (%)	Molefraction in % for 54 steps				
		90	98	99	99.5	99.9
55		0.3	33.6	58.1	76.3	94.7
54		2.0	37.0	31.7	20.7	5.1
53		6.0	20.0	8.5	2.8	0.1
52		11.5	7.1	1.5	0.2	
51		16.3	1.8	0.2		
50		18.1	0.4			
49		16.4	0.1			
48		12.5				
47		8.2				
46		4.6				
45		2.3				
44		1.0				
43		0.4				
42		0.1				

187 coupling steps; e.g., human growth hormone

Number of amino acids	Coupling yield (%)	Molefraction in % for 187 steps			
		98	99	99.5	99.9
188		2.3	15.3	39.2	82.9
187		8.7	28.8	36.8	15.5
186		16.6	27.1	17.2	1.4
185		20.8	16.9	5.3	0.1
184		19.6	7.8	1.2	
183		14.6	2.9	0.2	
182		9.0	0.9		
181		4.8	0.2		
180		2.2	0.1		
179		0.9			
178		0.3			
177		0.1			

and fragments of the above mentioned size, if one utilizes its principles with the necessary critical care.

In this region of molecular weights of target peptides, the purification problems are comparable to those of conventional synthesis. See, for example, the excellent classical synthesis of secretin by E. Wünsch and co-workers [56] and their extremely tedious procedure of purification to obtain the completely active hormone [57]. Comparable efforts were necessary to isolate small amounts of active synthetic ACTH indistinguishable to the natural hormone from an end product mixture of a stepwise solid phase synthesis by C. H. Li and collaborators [58].

On the one hand, the chance of purifying an inhomogeneous material depends upon the relative differences in the molecular weights and ionic charges between the main product and contaminations. On the other hand, the statistical distribution of a failure along the course of a solid phase synthesis results in two types of defective sequences, which together form the binominial combination of the composition of by-products, mentioned above. Quoting from E. Bayer and co-workers (1969) [54] "we can distinguish between the truncated sequences, in which amino acids are missing from the amino end, and the failure sequences, in which amino acids are missing from within the chain". In Figure 15 it is demonstrated that the bulk of by-products closely related to the target molecule, not only in terms of molecular weight, is represented by failure sequences, whereas the truncated sequences differ more pronouncedly from the main product and are formed theoretically in a smaller amount. The chance and success of the Merrifield synthesis of smaller peptides are founded on this small difference in possible by-products (again citing E. Bayer, 1970 [54]), "if a truncated sequence cannot couple in later steps, no failure sequences are present, and the isolation of the desired peptide is much more likely to be achieved than in the case when all possible failure sequences are formed".

From my point of view of practical use of the Merrifield synthesis, this reason has to be judged as a "categorical imperative" of the solid phase methodology. As long as complete peptide elongations on solid phase cannot be guaranteed by any analytical procedure, one has to suppress the statistical formation of failure sequences as far as possible by trying to block the truncated ones from further reactions. The chemical realization of this methodical necessity described in Sect. 3.4 to date is a question of routine and should no longer be a matter for sophisticated discussions. Even incomplete blocking of truncated sequences reduces drastically the formation of by-products and facilitates the efforts for purification (for further explanations see p. 60).

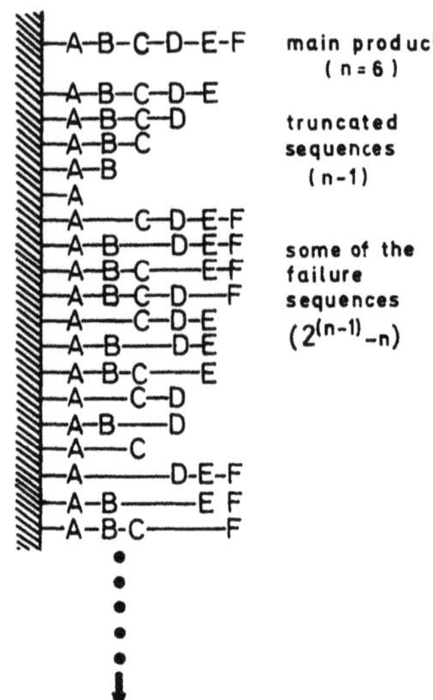

Fig. 15. The schematic description of "truncated" and "failure" sequences in the Merrifield peptide synthesis characterizing the product composition

Though quantitative conversions on solid phase are hard to reach, besides the imperative to block truncated sequences there is another excellent chance to limit the theoretical microheterogeneity of the end product of a Merrifield synthesis. As shown above, the distribution and quantity of contaminations increase exponentially with the number of stages of an incremental synthesis, whereas the possibility of distinguishing between the main and by-products decreases in the same exponential proportion.

This situation changes considerably if one conducts even an extended stepwise peptide synthesis on solid phase with small peptide units for building blocks instead of amino acid derivatives. In this way, the number of stages to synthesize a polypeptide is diminished to a fraction of those necessary in the original Merrifield procedure, with all the advantageous consequences of a limited statistical expansion of failures along the synthesis run.

Granted that in this modified form of a Merrifield synthesis the probability of quantitative N-terminal fragment condensations on solid phase is further reduced by the size of the building blocks, even these enlarged units of molecular weight render a facilitated end product purification possible especially in combination with the use of any blocking of truncated fragments. Later, this very promising variant for the future development of combinations between the conventional and solid phase concept will be outlined in more detail.

3 Chemical Details of the Method

The selection of methodical procedures noticed in this chapter should not constitute a devaluation of the scientific-intellectual wealth of the approaches and modifications omitted here, which often are chemically fascinating but less generally applicable from the point of view of practical preparative use. For the same reason, several details of the original method will be mentioned only in passing, since some of them became obsolete by later discovered aspects.

3.1 The Polymer Support, Its Choice, Properties, and Preparation

The rules for the choice of a polymer suitable for solid phase peptide syntheses are characterized by the demand for
a) **macroscopic insolubility**
b) **mechanical stability,** and for
c) **chemical inertness** of the nonfunctionalized principal support under all conditions of the repeated operations of the Merrifield synthesis.
d) **The chemical constitution** of the polymer has to render both the covalent binding of an amino acid, in a stage preceding the solid phase peptide synthesis and the cleavage of the peptide end product from the support.
e) **The physical and chemical properties** of the polymer have to favor chemical reactions at aminoacyl sites on polymer and have to facilitate all wash-out operations during solid phase synthesis.

In the incipience of the methodical development, after some tests with different types of supports, Merrifield had already selected from the gigantic palette of organic and inorganic polymers the commercially available beaded polystyrene (200–400 mesh, 80–20 μ diameter), cross-linked by 2% divinylbenzene as the most suitable up to that time for the purposes of peptide synthesis on solid phase. Today there are series of investigations to find better supports for use in Merrifield's synthesis (for comparison, see the review articles of Merrifield [16, 35] and Meienhofer [33]). But in all cases the improved properties of novel carriers or modified polystyrenes concern only one or two of the above-mentioned necessary parameters – e.g., mechanical stability or strengthened C-terminal bond of an amino acid to the carrier – whereas nearly all the other characteristics for a suitable solid phase turn out to be less favourable, compared to the original Merrifield resin.

The particular qualification of cross-linked polystyrene in the Merrifield synthesis can be explained by the following facts:

Copolymerizations of styrene and divinylbenzene (DVB) in suspension yield the polymers in small spherical particles, which are insoluble in all known organic solvents down to degrees of cross-linkage as low as about 0.2% of DVB. This fulfills not only the first rule for the choice of a suitable polymer (even with 0.1% DVB only 5% of the beaded copolymer can be dissolved with toluene [59]) but favors wash-out processes by the outer shape of the round particles (demand 5). The chemical-constitutional uniformity of the beads, which consist of ethylenebenzene units only, bridged on few points by diethylenebenzenes, guarantees, on the one hand, the chemical inertness to all conditions of the Merrifield peptide synthesis, but favors, on the other hand, the functionalization of the support by a variety of one-step aromatic substitution reactions preceding the peptide synthesis (demands 3 and 4). All these aspects are extraordinarily supported by the gel state of low cross-linked polystyrenes in appropriate solvents like hydrocarbon-halides, dimethylformamide, cyclic ethers, and especially toluene (demand 5). Depending on the solvation power of these solvents to the macroscopically insoluble macromolecular coil — cross-linked at very few points — the beads swell to a considerable volume the smaller the number of polymer threads fixed by cross-links (Table 3). In this gel state, the solvated coil exists in the state of dissolved macromolecules fixed in space with relatively free movement for quite larger segments of the polymer chain in the "included" solvent. The more a gel is swollen, the greater is the freedom of movement since the coil density decreases at a given degree of cross-linkage with the increasing volume of the "included" solvent (gain of entropy). The chain segments of a low cross-linked coil matrix are in continuous motion (intramolecular Brownian motion). Quoting from B. Vollmert [60], "through the motion the gel-coil undergoes a constant intensive blending action, so that the reactivity of the chain molecule

Table 3. Solvation of cross-linked polystyrene gels

Degree of cross-linkage[a]	2%	1%	ca. 0.3%
Solvent[c]	Swelling Factors[b]		
Trichlormethane	3.2	4.3	13.4
Tetrahydrofuran	3.7	5.2	12.3
Toluol	2.5	4.7	12.5
Pyridin	2.4	4.2	11.5
Dichlormethane	2.5	4.3	11.0
Dioxan	2.6	3.5	10.7
Dimethylformamide	1.3	2.6	6.8
Ethylacetate	1.6	2.1	5.5
tert. Butanol	1.0	1.1	2.2
Methanol	1.0	1.0	1.8

[a] Divinylbenzene isomers per weight of copolymer.
[b] Determined by gain in weight by solvent/g of dry beads, reduced to volume of solvent/g dry beads; 50 hours swelling, 20 °C.
[c] According to decreasing solvation power.

is not considerably reduced in any way by the gel character of the coil. This explains why cross-linked gels ... are still highly active reaction media ... This is particularly important for the reaction of living organisms, where nearly all reactions in the cell occur in a gel state". The quasidissolved state of a swollen gel is mainly determined by the cross-link density of the polymer. Yet the free flexibility of the solvated threads of the macromolecule are restricted to a small extent by the constitution of the polymer. By its contents of aromatic side groups, the coil of styrene-DVB copolymer has a somewhat more rigid structure, or in other words, a higher degree of orientation in the included solvent than hydrocarbonpolymers like polyethylene, for example. For these reasons styrene-copolymers with less than 1% of divinylbenzene are found to be nearly ideal supports for the purposes of the Merrifield synthesis. The only remaining question is the mechanical stability of more or less gelatinous particles (demand 2). The lower the degree of cross-links on the one hand, the higher the quasidissolved state of the gel and therefore its reactivity can be considered. But, on the other hand, one can assume that the more a gel is swollen, the lower its mechanical stability. Beads of polystyrene with 2% DVB, swollen in toluene, still feel "hard" between your fingers, whereas 1% cross-linked particles feel "soft" and compressible. Swollen beads of polystyrene, with contents of less than 0.5% divinylbenzene for cross-linkage, look gelatinous and glassy like frog spawn, and can be easily squashed between the fingers. (This type of polymer today is mainly used in the author's laboratory.) The utilization of the highly swollen, quasidissolved − and therefore favorable − state of low cross-linked spherical gel beads is limited only by the way the gel particles are mixed and agitated with reagent solutions and washing media. Unfortunately, the various types of machines for the purpose of more or less automatic Merrifield peptide synthesis (see p. 74) prevent the use of 0.5−0.2% cross-linked gel supports because of lack of equipment from reactors, in which gel and liquids are mixed by shaking, stirring, or even by vibrating. To separate the phases, the support is filtered off on relatively small porous glass filter disks, which act like a rasp on soft beads during agitation and would be blocked rapidly by ground gel particles. For these reasons a novel reactor was constructed [61], which allows the intense penetration of gelatinous beads as mechanically labile as 0.2% cross-linked polystyrene − e.g., in the Merrifield synthesis − without destruction of soft particles (for details see p. 75).

To sum up, at first it should be noted that the Merrifield peptide synthesis on supports like 1% cross-linked polystyrene does not deal with solid phase reactions, even if the particles look solid. Secondly, we have to imagine that the free motion of large chain segments in the maximum swollen state of a macromolecular gel phase − which is cross-linked to such a low degree, that it is just insoluble macroscopically − guarantees the reactivity of the same non-cross-linked, soluble macromolecule. Therefore, Merrifield syntheses should be performed on styrene copolymers of the lowest possible degree of cross-linkage. There does not exist a sharp demarcation between peptide synthesis on "solid" gel phase and so-called "liquid" phase [62] from the point of view of behavior and reactivity of the macromolecules in concentrated solutions, which already have gel characteristics.

On the contrary, investigations, which dealt with the opposite intention, to utilize the hard structure and mechanical advantages of highly porous macroreticular resins (cross-linked to > 55% DVB), or even of glass beads for true *solid* phase peptide synthesis through surface reactions, proved to be of scientific interest only, but were unsuited for preparative peptide synthesis. Even modifications of the different types of solids by "handle" or

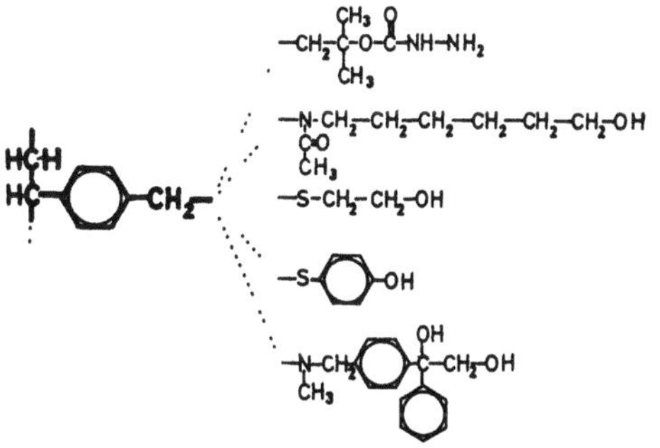

Fig. 16. Several types of spacer molecules used
in solid phase modifications

spacer molecules (Fig. 16) did not change the negative results in general, as we learned
from our own results. Drastic conditions are necessary to functionalize macroporous polystyrene beads to reach a capacity of reactive sites of $0.01-1$ μ equiv./g solid, which unfortunately are not completely convertible by amino acid residues. Even the peptide syntheses
proved to proceed with lower reaction rates and yielded a quantity of the end product
more than a thousand times lower compared to syntheses on gel type supports.

In the following paragraph a description is given for the preparation of very low cross-
linked polystyrene gel, which can hardly be obtained on the market.

Suspension Copolymerization of Styrene with $\lesssim 0.5\%$ Divinylbenzene. In a 1 liter three-
necked round-bottom flask with reflux condenser, under a slow stream of nitrogen (inlet
above the liquid surface), 800 ml of deionized water is warmed up to 50 °C in a silicon oil
bath (3 liter, 500 watt heater), which covers the lower half of the reaction vessel. In the
middle neck, a glass stirrer (ending 2 cm above the bottom) of the shape and dimensions
shown in Figure 17, is set in with a teflon holder for high speed stirring. With gentle stirring, 3 g of pulverized Moviol N70/88 (polyvinylalcohol, Farbwerke Hoechst, Frankfurt/M.,
Germany) is dissolved within 2 hours. 200 g styrene and 2 g divinylbenzene (containing
about 50% of divinylbenzene isomers and 50% ethyl-styrene isomers) — destabilized by
extraction with 2 N NaOH twice before use — is added to the aqueous phase. The speed
of the stirrer now is set on $1,800-2,000$ rpm and must remain there without any interruption during the whole procedure (15 hours). 5 min later, 1 g of benzoyl peroxid, as a
wet powder, is added to the emulsified monomers. Now, the temperature of the oil bath
has to be raised within 60 min to 90 °C (no thermometer in the reaction vessel!) and is
kept at this temperature for 15 hours of stirring. The polymerization becomes critical in
the third hour of heating (adhesive rubber consistency of the copolymer pellets). At the
end of the reaction time, under stirring, the suspension is cooled to room temperature.
The polymer beads are collected on a porcelain funnel (ϕ 40 cm), and washed with water
and methanol. The air-dried pellets, in appropriate charges, are transferred to a soxhlet
apparatus and extracted for 72 hours with refluxing dioxane, which contains some pieces
of sodium metal for binding of water and peroxides.

The dioxane-swollen purified gel polymer beads (ca. 500 ml) are transferred in small
portions into 2,000 ml of gently stirred pure methanol for shrinking of the pellets, which
finally are collected on a funnel, washed with pure methanol, and dried in vacuo for 24
hours at 40 °C. The dry material is sized on a sieve-pack yielding ca. 80 g (40%) of poly-

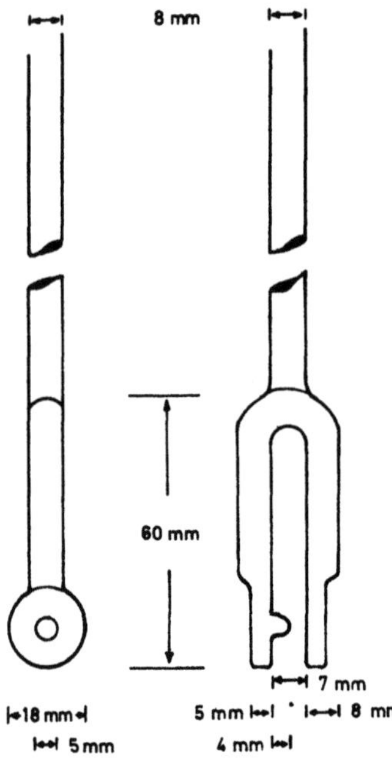

Fig. 17. The shape and dimensions of a glass stirrer used in the suspension copolymerization of styrene and divinylbenzene to prepare a low cross-linked gel phase

mer beads of 60—115 mesh diamter. To prepare suspension polymers from styrene-divinylbenzene with 1—2% cross-linkage, the unmodified and shorter procedure of Brown, Cherdron and Kern [63] is recommended.

Recently, a cross-linked polymeric support of the gel type was prepared, which seems to overcome a still existing general problem of peptide synthesis on cross-linked polystyrene, namely the different solvation behaviour of the growing polar peptide chain and the lipophilic styrene matrix.

Sheppard and co-workers [64] copolymerized the ternary system consisting of N,N-dimethylacrylamide (the macromolecular thread forming monomer), N,N'-bisacryloylethylenediamine (the cross-linking agent) and N-tert.butoxycarbonyl-β-alanyl-N'-acryloylhexamethylenediamine (the monomeric component, which introduces functional sites into the copolymerizing gel; for detailed discussion see the next section).

The cross-linked polar gel so constructed fulfills all requirements characterizing a suitable support for the Merrifield synthesis (see p. 16), combined with the great advantage of comparable solvation properties of the support and the peptides. The modified polyarylamide gel swells in dimethylformamide and acetic acid to ten times its dry volume, even more than in water, but very much less in dichloromethane and other less polar organic solvents. These properties are the reverse of those of polymeric supports based on styrene. The first examinations of this new type [65, 205, 206] of polymer by Merrifield syntheses demonstrate the powerful potential of this support even for investigations in which those synthetic peptide sequences were directly analyzed on the gel by Edman degradation [66]. The similarity in the solvation properties of the peptide and the polyamide support to which it is attached favors the speed and completion of all reactions during synthesis and degradation as well.

3.1.1 Introduction and Conversion of Functional Sites on Polymer

The uniform constitution of polystyrene from ethylenebenzene units characterizes the selection of chemical reactions to introduce functional groups into the aromatic rings. For this purpose, mainly Friedel-Crafts alkylations and acylations are utilized. From the author's experience, however, in publications dealing with modifications of polystyrene, it is not sufficiently emphasized that only single stage conversions on an insoluble polymer may result in an accurately defined functionalization of the support. Depending on the degree of cross-linkage, which determines the heterogeneity of the reaction centers in the matrix, there remains from each additional transformation on polymer sites a small quantity of nonconverted functions of the pre-stage, so that a support which was modified in several steps finally contains small portions of undesired groups of diverse reactivity (example in Fig. 18) besides the desired functions. This small amount of reactive sites from pre-stages, however, is not detected by IR, since those bands of low intensity in IR spectra of insoluble polymers are hidden behind the background noise of the recording.

Nevertheless, such reactive functions from pre-stages on a modified support may interfere later on with steps of the Merrifield peptide synthesis causing unforeseen troubles. From this shortly depicted complex of problems of subsequent modification of polystyrene supports, the following consequences may be deduced:
a) For best defined functionalization of polymer supports only single stage reactions are recommended, e.g., chloromethylation [67] and acylation with bromoacetyl bromide [68, 69].
b) Modifications which need several stages of preparation should be performed on the monomeric precursors, which in the final form may be converted to vinyl compounds and copolymerized [64].
c) In cases of polystyrene supports, the second demand usually is hard to realize because of the incompatibility of vinylbenzene and Friedel-Craft reactions for ring substitution. Therefore the only way to remedy this problem is to affix specially constructed molecules, which contain the functions and properties desired (so called spacer and handle molecules), to chloromethylated polystyrene supports (Fig. 16).

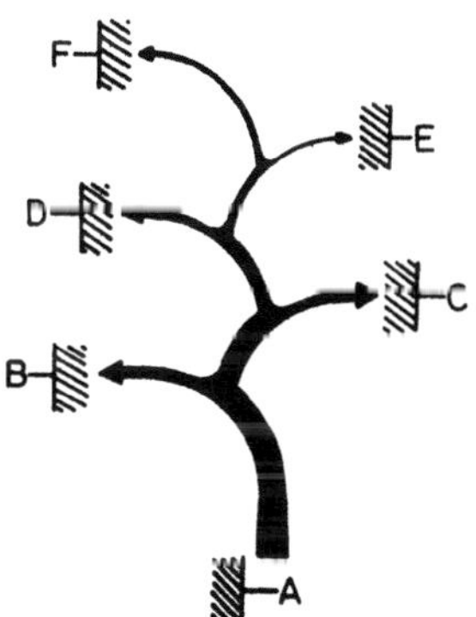

Fig. 18. The multifunctionality of cross-linked polymers after several synthetic transformations to introduce a distinctly reactive site into the support

From the point of view of practical utilization, only the chloromethylation of polystyrene with chloromethyl-methyl ether has, to date, attained relevance in functionalization of Merrifield supports.

Frank and Hagenmaier [70] in a detailed study investigated the influence of various chloromethylation procedures and varying degrees of cross-linkage of polystyrene on the course of the Merrifield synthesis of a model pentapeptide, performed on the supports.

Table 4 very clearly verifies our lengthy experience that the best synthetic results are gained on polymers which are cross-linked to the lowest possible degree. Because the interplay between the swelling properties of a support and the type of chloromethylation procedures has an influence on the success of Merrifield syntheses, these correlations will be discussed in more detail in the light of general aspects on functionalization regarding the chloromethylation as a model reaction.

If chloromethyl methyl ether/$SnCl_4$ is added to 1% cross-linked polystyrene beads, preswollen in chloroform, a radial gradient of the dissolved Friedel-Craft complex will result in the "included" solvent of the relatively free moving solvated macromolecule. The concentration of the reactants is low at the statistically distributed, relatively easily accessible phenyl moities of the swollen polymer, since the large volume of the "included" solvent (depending on solvation power of chloroform and degree of cross-linkage) acts as a very diluting medium. A uniform but low chloromethylation of the highly swollen polymer support can be expected. If this chloromethylated polymer now is reacted with a N-protected amino acid in a solvent like ethyl acetate [70], which has half of the solvation power of chloroform (see Table 4), parts of the functional sites will not be transformed into amino acid ester links, since the polystyrene threads in ethyl acetate are not solvated to the same extent as formerly was the case in chloroform. Therefore chloromethyl functions will remain inaccessible. A rate of amino acid ester formation on a polymer smaller

Table 4. The influence of chloromethylation procedures and of the degree of cross-linkage on the functionalization of polystyrenes, according to Frank and Hagenmaier [70]

Procedure[a]	Cross-linkage [% divinylbenzene]	Chloromethyl functions [mMols/gr. polymer]	Load[a] [mMols proline/gr. support]	Swelling factor[d]
1[b, e]	2	4.1	0.83	4.0
1[e]	2	3.8	0.50	3.8
3[b, e]	2	4.5	0.78	4.0
1	1	2.3	0.85	6.0
1[c]	1	0.8	0.10	5.9
2[c]	2	1.4	0.32	4.0
4[c]	2	1.4	0.22	4.0

[a] Preparative details are given in [70].
[b] 1 = [7.5 ml $SnCl_4$ in 50 ml $ClCH_2OCH_3$] per 50 grs. of polystyrene in 300 ml solvent.
3 = 3 x [ ] per 50 grs. of polystyrene in 300 ml solvent.
[c] In chloroform.
[d] Determined by gain in weight and reduced to volume of dichloromethane/gr. of loaded support.
[e] In hexane.

than the original chloromethyl capacity can be detected. These N-protected amino acid sites, on the other hand, are most easily transformed during subsequent peptide synthesis, since they are located in best solvated areas of the cross-linked macromolecular support.

The situation is somewhat different in the case of chloromethylation of the same type of low cross-linked polymer in its shrunken state (nonsolvated) in n-hexane. This solvent coats but does not solvate the polymer beads. When the reactive complex of chloromethyl methyl ether/stannic tetrachloride is added, the macromolecules on the surface zone of the polymer pellets are immediately solvated due to much higher concentration of reactants at reactive phenyl moities, in contrast to the highly diluted state of the Friedel-Crafts alkylation in the first example. On the other hand, the relatively small amount of chloro-methyl-methylether/$SnCl_4$ added in n-hexane is only sufficient to solvate and function-alize most exposed outer parts of the macromolecular thread in the shell zone of the poly-mer, depending on the accidental distribution of cross-links in that area of the macromole-cule.

As the functional sites introduced in this way are located in positions which even in the shrunken state of the support are easily accessible, this type of chloromethylated poly-styrene exhibits excellent properties for further transformations in the Merrifield peptide synthesis. Nevertheless, we should retain that this type of reactive polymer may only con-tain its chloromethyl functions in a relatively dense filling in the shell zone of the polysty-rene particles.

A uniform chloromethylation mainly selecting all exposed positions in the whole macromolecular matrix is attained in the quasidissolved state of low cross-linked ($< 0.5\%$ DVB) polystyrene by preswelling the support in pure chloromethyl-methyl ether (swelling factor 12; maximal solvation of the polymer) and subsequent starting of the alkylation by addition of small quantities of $SnCl_4$ at 0 °C in a short time reaction (15 min). Under these conditions the reactive complex of chloromethyl-methyl ether/$SnCl_4$ will attack mainly the most easily accessible phenyl moities all over the polystyrene matrix. The loading of the support with chloromethyl functions can be controlled very conveniently by the con-centration of $SnCl_4$ added [71] (Table 5). Recently the use of $ZnCl_2$ as a less active Frie-del-Crafts catalyst was recommended for the same purpose [72].

Chloromethylation of $< 0.5\%$ DVB-polystyrene gel [71]. With gentle stirring, 10 g of dry support are swollen for at least 1 hour in 200 ml of freshly distilled chloromethyl-methyl

Table 5. Rate of chloromethylation depending on concentration of catalyst

$SnCl_4$[a] (ml)	Chloromethyl capacity[b] (m equiv./g[c])
1	0.89−1.00
0.5	0.26−0.3
0.1	0.1−0.15

[a] Reaction conditions given above.
[b] Determined by Carius method.
[c] $< 0.5\%$ cross-linked (DVB) polystyrene gel.

ether at room temperature. After cooling of the suspension to exactly 0 °C, 0.1–1.0 ml of $SnCl_4$ are added under vigorous stirring. Agitation then is slowed down again and continued for exactly 15 minutes at 0 °C. On a glass filter funnel, the chloromethylated polymer is quickly separated from the reaction liquid, which is directly transferred into a distilling apparatus for recovery of excessive chloromethyl-methyl ether.

The functionalized support is immediately washed five times with 100 ml each of dioxane, dioxane/2nHCl (3:1, v; v), and dioxane/water (3:1, v; v) on the glass filter funnel to remove the remaining chloromethyl-methyl ether/$SnCl_4$ ($AgNO_3$-test on chloride ions in the last filtrate). Thereafter, water is displaced from the chloromethylated polystyrene five times, with 100 ml each of dioxane, dioxane/methanol (1:1, v; v), and methanol, which gradually shrinks the gel polymer. The air-dried material finally is dried in vacuo in the presence of phosphorus pentoxide for 15 h at 50 °C.

The various types of chloromethylation were described to illustrate that for transformations on insoluble but swelling polymers the crucial factor generally does not mean the concentration of reactants in the "free" liquid phase, but the concentration of reactive components at the points of reaction in the matrix. This is explained in the following example. Let us assume a reaction vessel of any type is charged with 1 g of polymer which has a capacity of reactive functions of 0.5 mmole. A soluble compound, dissolved in 20 ml of dichloromethane is added in threefold molar excess to transform the functional sites on the polymer as completely as possible. The swelling of the low cross-linked support in dichloromethane, however, has the factor of 10, which means that the inner volume of 1 g of the polymer, solvated by the dichloromethane solution, is 10 ml. Therefore the effective concentration of the dissolved excess component at the reactive sites has only half of the desired strength.

This "effective" concentration is determined by the degree of solvation of the cross-linked macromolecule in the liquid reaction mixture and by the time for diffusion of reactants into this inner volume of the polymer. Since in particles cross-linked to $\leq$ 1% (100–400 mesh) diffusion requires about 1–2 minutes, this parameter might be neglected when judging possible conversions on the gel phase, whereas it is essential to know the swelling factor of the polymer in the specific mixture of solvent and reactants. From this point of view, in the example of chloromethylation in n-hexane, the state of reaction of the polymer cannot be described exactly, since the small amount of solvating reagent just may suffice to transform outer parts of the coiled macromolecule into the reactive state.

After having discussed these principal aspects, some modifications on polystyrene are described in the following to punctuate the limitations mentioned above. N-protected amino acids, attached onto chloromethylated polystyrene, form links of the benzyl ester [73] type well known in conventional peptide synthesis (for details see the next section).

By the standard procedure to cleave peptides from this polymer benzyl ester, the synthetized sequences are released completely deprotected (see Sect. 3.5) and therefore are not best suited for further fragment condensations. This fact forced some research groups to develop modified polystyrenes, which incorporated a higher selectivity of bond strength between C-terminal anchor linkages of peptides to the functional sites on the polymer and to protecting groups on side functions of the growing peptides. (More detailed aspects on choice of protecting groups are discussed in Sect. 3.2.1.) This would render the possibility of cleaving peptides from their support without loss of protecting groups, and of combin-

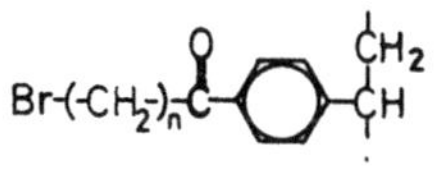

Fig. 19. Haloacylated polystyrenes

ing the protected sequences in subsequent fragment condensations, e.g., in solution or on polymer phase. To reach this goal, various modifications of the binding function on a polystyrene support were proposed and fully reviewed by Merrifield [16] and Meienhofer [33].

As another example for one-step functionalization on polystyrene, the Friedel-Crafts acylation with haloacetyl and -propionyl bromide should be mentioned [68, 69] (Fig. 19). The haloacylation is performed in nitrobenzene — which has the swelling power of toluene (see Table 3) — catalyzed by aluminum chloride at room temperature for 20 hours, yielding a capacity of 2.2 mequiv. of reactive sites/g of 2% cross-linked polystyrene [69]. The loading with bromoacetyl moieties of 0.05 mequiv./g found in another experiment [68] should be regarded as the lower limit of functionalization for practical use in the Merrifield synthesis.

The haloacetophenone type of polymer reacted with N-protected amino acids and peptides — as already known from esterifications with the monomer reagent in conventional syntheses [74] — under milder conditions but in better loading yield compared to chloromethylated polystyrene, resulting in enhanced reactivity of the peptide phenacyl ester bond on polymer towards nucleophilic cleavage reagents [74—76]. (For further details on peptide cleavage, see Sect. 3.5.)

The electrophilicity of the acetophenone carbonyl in haloacylated polystyrene may cause side reactions like Schiff base formations (Fig. 20), and the acylation potency of the phenacyl ester grouping may tend to form diketopiperazines on the dipeptide stage of a Merrifield synthesis (Fig. 21). However, this electrophilicity, on the other hand, characterizes the reactivity of the haloacyl function to bind not only N-protected amino acids, but also peptides, onto the support, and solidifies the binding function between peptide and polymer support against the attack of acid during peptide synthesis. Therefore, to balance advantages and shortcomings of haloacylated polystyrene against another, more experience with this type of functionalization would be necessary than published up to now.

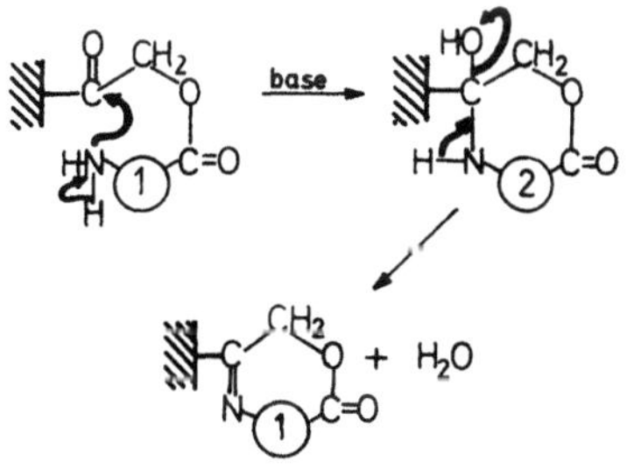

Fig. 20. Possible Schiff base formation on peptides bound to polymer carriers via phenacyl ester bonds

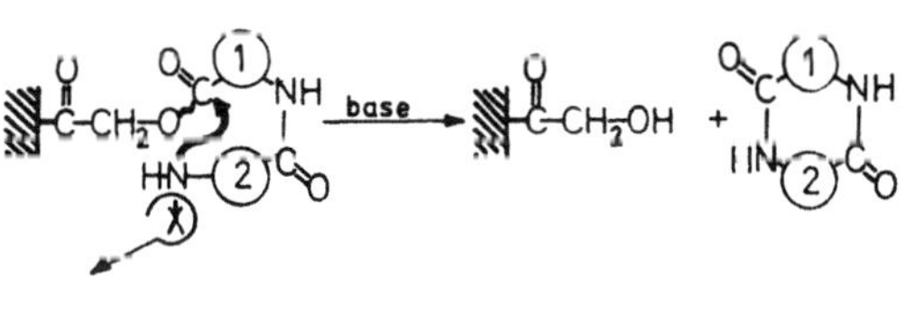

Fig. 21. Possible loss of dipeptides via dioxo piperazine release from phenacyl links in polymer support

25

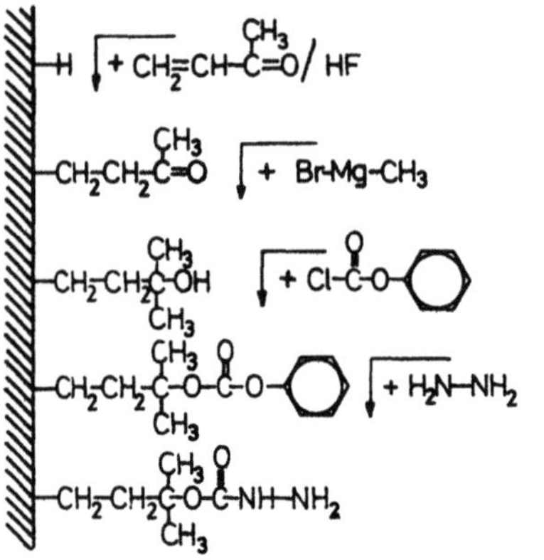

Fig. 22. The preparation of tert. alkoxycarbonyl hydrazide sites on cross-linked polystyrene

With regard to subsequent fragment condensation, it is necessary to preserve the side chain protecting group of a peptide during cleavage from its polymer support. Having this in mind, Wang and Merrifield [77] very elegantly functionalized the benzene ring moieties of polystyrene by tert. alkoxycarbonyl hydrazide sites in four stages, handling the 2% cross-linked support like a conventional chemical compound (Fig. 22).

This thoroughly completed preparation is taken as an example in discussing some problems deriving from multistage conversions on an insoluble polymer, which do not refer to the suitable labile attachment of peptides to this modified support discussed in Sect. 3.3.5.

The first conversion on a polymer to introduce a ethyl methyl ketone moiety was performed, without dilution, in the liquid reactants vinyl methyl ketone and condensed hydrogen fluoride, which scarcely solvate the polystyrene support. Recently, Tesser and co-workers [78] reported an unsuccessful attempt to verify this type of transformation on polystyrene, supposing that their failure depended upon the particular batch of polymer used. Obviously the swelling properties of the cross-linked supports in the poorly solvating reaction liquid — condensed hydrogen fluoride — differed in these independent experiments. Therefore, to describe completely a procedure to modify an insoluble, but swelling, polymer, it is essential to specify the solvation power of the liquid mixture of reactants.

The introduction of ketone functions in the above-mentioned modification experiment, was determined qualitatively by its characteristic absorption at 1,725 cm^{-1} in the IR spectrum. This functional group in the second stage was converted to a tertiary alcohol by reaction with methyl magnesium bromide on solid phase. The material obtained was reported to show no carbonyl absorption; in other words, complete conversion of the carbonyl function on polymer was deduced from the IR spectrum, whereas this type of Grignard reaction in conventional syntheses in real solutions is known to proceed with 80–90% yield [79]. From the microanalysis of the modified polystyrene, 2.4 mmoles of oxygen containing functional sites can be calculated, from which in a third stage parts of the tertiary alcohol groups were reacted with phenyl chloroformate to form mixed carbonate functions. Their hydrazinolysis in the last stage yielded 0.41 mmole of tert. alkoxycarbonylhydrazide/gram of polymer, suitable for the attachment of protected amino acids and for acid-sensitive cleavage of peptide hydrazides from this support without loss of protecting groups on side functions.

Granted that each stage in both this modification procedure and in the verifiable procedure of Tesser and co-workers [78] was carefully elaborated and optimized, nevertheless, in terms of conventional synthesis it would be misleading to consider this modified polymeric support a uniform chemical compound. We can estimate that the insoluble macromolecule besides its tert. alkoxycarbonylhydrazide functions (0.41 mmole/g) may contain about 0.2 mmole/g of 3-oxobutyl residues (ca. 10% of the total oxygen containing sites; limit of the accuracy of identification in the IR spectrum on polymer; 90% conversion in the Grignard reaction in heterogeneous phase). About 0.4 mmole/g of tert. alcohol group may have remained unreacted, since from our experience their transformation into mixed phenyl carbonates — even with a threefold excess of phenyl chloroformate — on solid phase scarcely reaches 80%.

The final hydrazinolysis of the carbonate sites on polymer yielded 0.41 mmole/g tert. alkoxycarbonylhydrazide, as already mentioned. Based upon the estimations given above, this means about 70% or 1 mmole/g of the carbonate functions (1.4 mmole/g) have remained unreacted.

To sum up, in this preparation, besides the desired hydrazide functions (0.41 mmole/g; ca. 20%, based on oxygen content of the first stage of preparation) there are about 1 mmole/g (ca. 50%, based upon oxygen contents of the first stage) of mixed phenyl-tert. butyl carbonate functions, 0.4 mmole/g (ca. 20%) of tert. alcohol groups, and 0.2 mmole/g (ca. 10%) of 3-oxo-butyl residues on polymer. In other words, very roughly estimated, about 80% of various functional sites with different reactivity form the reactive background of this polystyrene support, modified by tert. alkoxycarbonylhydrazide groups.

To evaluate the polyfunctional substitution more accurately, we may correct the estimation by using the analytical data given in the preparation verified by Tesser and co-workers [78]. In that procedure, the reaction of methyl vinyl ketone/liquid hydrogen fluoride to introduce 3-oxobutyl sites (first stage see Fig. 22) is evaded by the transformation of chloromethyl functions on polymer (0.64 mmole/g) with the sodium salt of tert. butyl acetoacetate carbanion which was reported to proceed completely, followed by acidic decarboxylation (Fig. 23). This preparation, in which the remaining steps are identical to those described by Wang and Merrifield, finally yields 0.43 mmole/g of tert. alkoxycarbonylhydrazide functions on polymer, with an overall yield of 71%. Thus the total contamination of the desired functional groups with the reactive sites from prestages makes up only 29%.

Granted that the polymeric support in this extended procedure on the 3-oxo-butyl stage still can be considered to be unifunctional, the 29% of undesired reactive groups derive from incomplete conversions on the keto, the alcoholic, and the carbonate stages. Using the estimation of the reaction rates as explained in the preceding example, we cal-

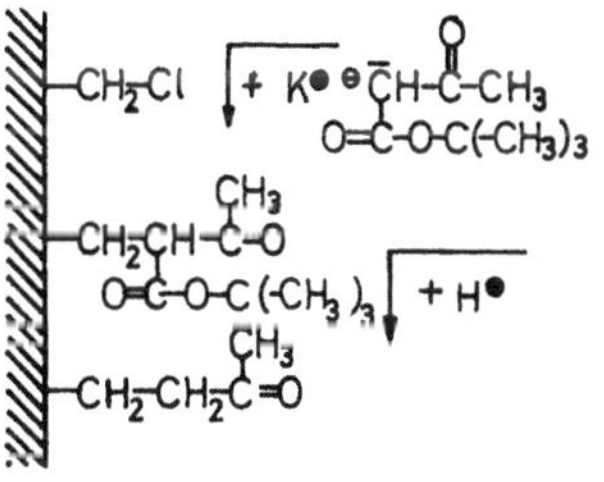

Fig. 23. The introduction of 3-oxobutyl sites into cross-linked polystyrene

culate 0.15 mmole/g of remaining mixed carbonate functions (25%, based on total content of carbonate functions, 23%, based upon the initial capacity of the support), whereas less than 0.1 mmole/g derives from the keto and the alcoholic stages of the preparation.

Though this multistage modification on polymer is performed with improved homogeneity compared to the first one, whose functional heterogeneity had been only very roughly estimated, both of the examples demonstrate that the finally modified support will under no circumstances exhibit functional uniformity. The possibilities of handling a cross-linked polymer like a conventional chemical in a projected pathway of synthetic operations are limited, since each deviation like side reactions and incomplete conversions remains fixed to the insoluble support. In addition, small quantities of functional sites (< 10%) are scarcely detectable by IR spectroscopy and are hard to analyze by conventional methods of organic chemistry.

With respect to the rules for the choice of suitable supports (see p. 16), a polymer containing different types of reactive functions can no longer be considered to be chemically inert to the conditions of the Merrifield synthesis, which are selected with regard to a specific functional group. The solvation properties of a multifunctional polymer differ from those of a uniformly functionalized one and complicate the estimation of suitable reaction conditions, but above all, the undesired functions may interfere during the varying steps and stages (see p. 6) of the peptide synthesis. Therefore we shall discuss this aspect in greater detail in the pertinent sections yet to come.

These objections generally can be raised on modifications of insoluble polymers due to the multifunctional nature of those supports provided with a distinct reactive site as a result of several stages of transformations on solid phase. The only alternative remaining is to prepare conventionally a definitely constructed reactive molecule that contains the desired function and to attach this compound in a single reaction to the polymer.

This way out also is the subject of numerous contributions in the literature reviewed [16, 33, 35]. The unique aim of all is the synthesis of spacers or handle molecules, inserted between the phenyl moieties of the cross-linked polystyrene thread and a reactive group, not only to attach protected amino acids suitable for peptide synthesis to the polymer, but to facilitate peptide cleavage without loss of protecting groups on side functions (see Fig. 16).

The so-called "safety-catch" concept (see also Sect. 3.5.3.1) combined with the utilization of specialized handle molecules on polymer, is demonstrated in the following example. In a conventional synthesis [80] 4'-aminomethyl-2,2-diphenyl-ethane-diol was prepared in six stages starting from p-aminoacetophenone (Fig. 24). The final glycol contains three functional sites of different reactivity. The very nucleophilic amino methyl group renders the possibility of anchoring the handle molecule to — e.g., chloromethylated — polystyrene. With the amino function in its protected form the primary glycolic site can simply be esterified by N-protected amino acids prior to the attachment of the spacer onto the polymeric phase. The tertiary alcoholic group is not subjected to esterification because of its sterical hindrance.

In this way, the esterification of an amino acid as well as the complete transformation of reactive functions, can be controlled by conventional analysis, independent from reaction conditions on polymeric phase. Furthermore, the problem of the fixation of a carboxylic compound to the polymer support is reduced to the smooth reaction of a primary amine to chloromethyl functions on cross-linked polystyrene, which were found to be com-

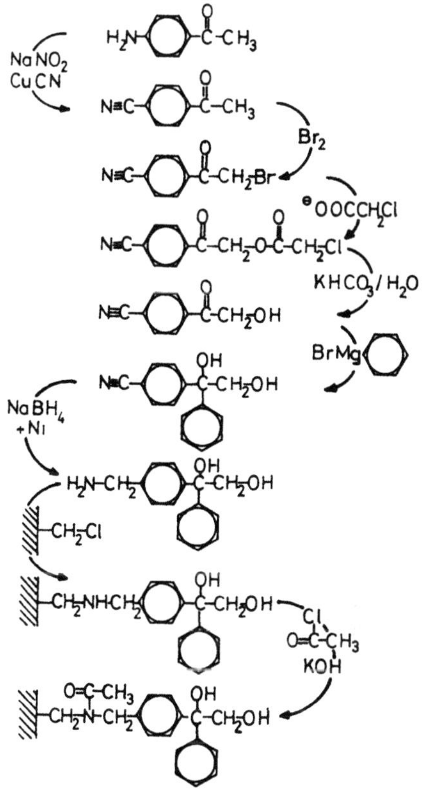

Fig. 24. The preparation of 4'-aminomethyl-2,2-di-
phenyl-ethanediol

pletely convertible by the handle amine up to chloromethyl capacities of 0.8—1.2 mmole/g of polymer, depending on the degree of cross-linkage. In extremely swelling polystyrenes (< 0.3% cross-links: quasidissolved state of the polymer) we noticed complete double alkylation of the handle (Fig. 25) amine by chloromethyl functions on polymer yielding tertiary amino groups of the spacer on support accompanied with additional cross-linkage of the gel phase, which subsequently shows a reduced swelling comparable to 1% crosslinked polystyrene [81]. To avoid this effect, the amino glycol molecule, prior to its literal utilization, generally was mono-methylated selectively on its nitrogen function.

It should be emphasized that the spacer bearing support modified in this way is indeed uniformly functionalized. The tertiary glycolic group was found entirely inert to the reaction conditions of the Merrifield synthesis, if low concentrations of trifluoroacetic acid (< 10%, in dichloromethane) are used in the deprotection procedures [81]. The subsequent elimination of water from the remaining tertiary alcoholic function of the glycolic handle which leads to the activation of the C-termini of peptides synthetized on this support, will be described in Sect. 3.5.3.1.

It should be mentioned that a spacer containing support, recently prepared by Rich and Gurwasa [82] has been reported to render the cleavage of fully protected peptide amides feasible by photolysis in methanol suspension of the polymer-supported peptide (Fig. 26).

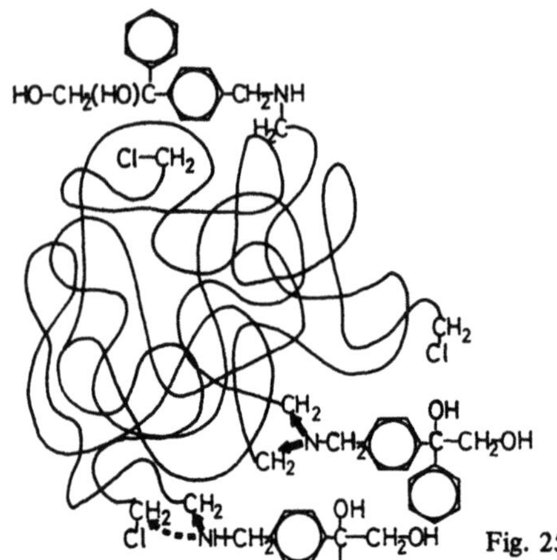

Fig. 25. The primary amino group of the spacer 4'-aminomethyl-2,2-diphenyl-ethanediol causes additional cross-linkage in chloromethylated polystyrene gel phase

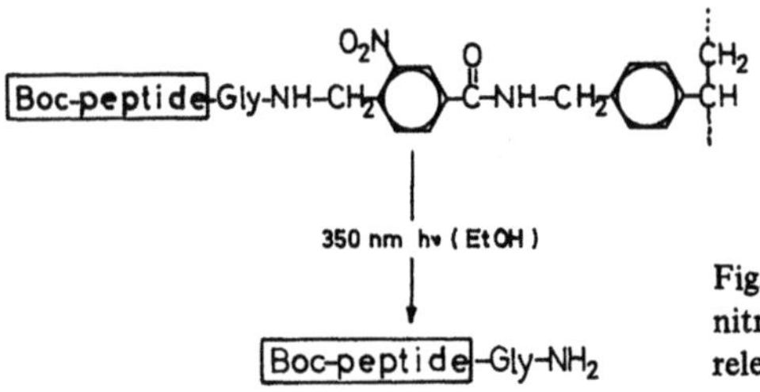

Fig. 26. The photolabile spacer acyl-4-aminomethyl-3-nitrobenzamidomethyl in polystyrene renders the release of peptide amides feasible

Finally, an example will be discussed in which a polymer support is uniformly functionalized by copolymerization of suitable monomers. As already mentioned in the beginning of this section, the monomers of polystyrene based supports — styrene and divinylbenzene — are not very suitable for functionalization prior to the copolymerization, since the reactivity of vinyl aromates is simply not compatible with reaction conditions required to functionalize the benzene ring by substitution reactions.

For reasons explained on p. 20 a modified N,N-dimethyl-polyacrylamide support was introduced [83] into the system of the Merrifield synthesis. This polar polymer is prepared by a ternary copolymerization of N,N-dimethyl-acrylamide, N,N'-bisacryloylethylenediamine and N-tert.butoxycarbonyl-β-alanyl-N'-acryloylhexamethylenediamine (Fig. 27). By the third component, functional groups are introduced into the gel polymer. After cleavage of the tert.butyloxycarbonyl protecting group the primary amino functions of the β-alanyl residues incorporated are liberated to form the anchor sites suitable to start a peptide synthesis or to react with further spacer molecules. Besides the advantageous solvation property of this type of polymer support, the great flexibility of the monomeric system to introduce a variety of functional moieties again punctuates the potential of this idea deviating from the polystyrene concept. Of further advantage is the low price and high reactivity of acryl chloride, the mother compound for preparing such polymers.

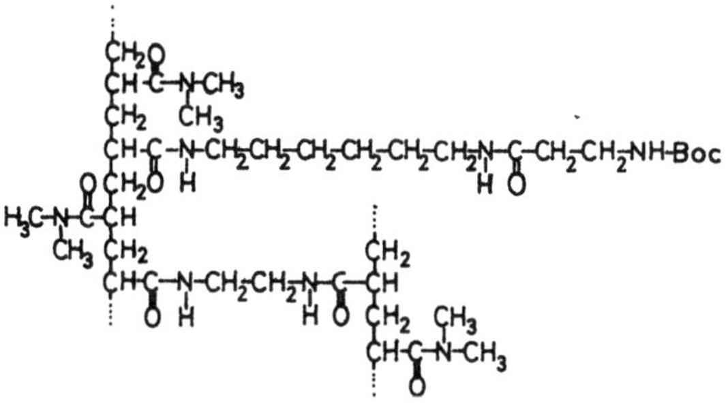

Fig. 27. The modified N,N-dimethyl-
polyacrylamide support for the purpose
of peptide synthesis in a polar microenvironment

To summarize, in several differing ways described above, functions like chloromethyl-, hydroxymethyl-, aminomethyl-, carboxymethyl-, bromoacetyl-, phenolic hydroxyl- and benzoate are mainly introduced into supports based on cross-linked polystyrenes.

3.1.2 Binding of Amino Acids and Peptides

The chemical reactions to conjugate amino acids or peptides with polystyrenes and other supports naturally depend on the type of functional sites on the gel phase. In most cases, chloromethyl groups are used to bind the starting N-protected amino acid — which is the C-terminal one of the target sequence to be synthesized — to the insoluble but swollen polymer. To date this is most efficiently done by utilizing caesium salts of the carboxylic components [84], which are simply prepared with caesium bicarbonate or hydroxide. To prevent base catalyzed racemization and side reactions on the polymer like saponification of ester bonds or of chloromethyl functions by caesium hydroxide not neutralized, N-protected amino acids are used in a slight excess.

Two milliequivalents of the lipophilic amino acid caesium salts, so formed in dimethylformamide, completely transform chloromethyl functions up to 0.8 milliequivalents/g of 0.5% cross-linked polystyrene at elevated temperature into benzyl ester bonds (Fig. 28) [85].

Fig. 28. The binding of N-protected amino acid caesium salts to chloromethyl polystyrene

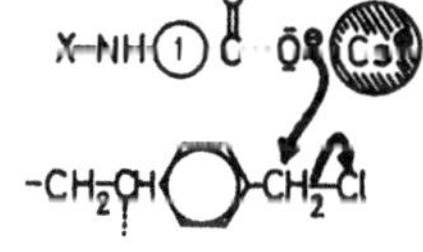

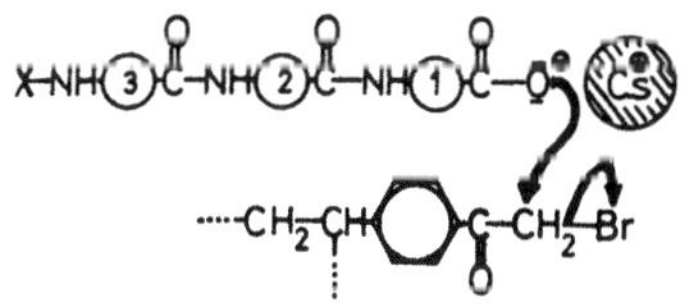

Fig. 29. The anchoring of N-protected tripeptide caesium salts
to bromoacetyl polystyrene

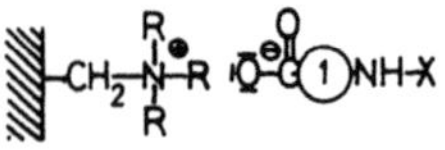

Fig. 30. The undesired ionic binding of N-protected amino acids to quarternary groups in polymer phase

However, already in a minor excess of 50% the caesium salts not only of N-protected amino acids but di- and tripeptides under identical conditions completely react with bromoacetyl groups on polymer, because of the enhanced electrophilic reactivity of those sites (Fig. 29) [85]. The phenacyl ester so formed are particularly suitable in experiments to synthesize fully protected peptide fragments, an aspect which will be discussed in Sect. 3.5.

A method as effective as the use of caesium carboxylates to react with chloromethyl and bromoacetyl sites on polymer is the utilization of amino acid tetraalkylammonium salts [86], because of the lipophilic character of the quaternary substituted cation, which in addition inhibits a side reaction possible with triethylammonium salts formerly used for the same purpose. Triethylamine reacts with chloromethyl and bromoacetyl groups forming quarternary centers on the polymer by which this partially becomes an ion exchange resin [67]. In a dual sense, those ionic functions would make the analytical control of the amount of covalently bound carboxylates by elemental analysis of the nitrogen content of the polymer doubtful: First, the ionic groups themselves on the polymer contain nitrogen and secondly, they could bind amino acids ionically. Both effects give the false impression of a higher loading of the support with amino acid than exists in reality (Fig. 30).

To circumvent this uncertainty without use of the caesium salt procedure, hydroxymethyl [87] or aminomethyl functions on the polymer can be reacted with carboxylic partners in an esterification reaction [207] or peptide bond formation with the aid of condensing agents like carbonylbisimidazole [88], dicyclohexylcarbodiimide [89] or others. For this purpose in the author's laboratory the use of symmetric anhydrides [3] of the N-protected amino acids to be attached to the support was found to be most effective [90], especially in the formation of activated esters on the gel phase with phenolic hydroxyl functions [38]. By this procedure, on 0.5% cross-linked polystyrenes, load levels up to 1.5 millimoles/g of the support are reached.

The subsequent binding of peptides, however, is not possible via symmetric anhydride formation. Even by dicyclohexylcarbodiimide and carbonylbisimidazol condensation with lipophilic tripeptides it is scarcely possible to bind more than 0.2 mmole/g of support. In those cases the attachment to bromoacetyl sites on polystyrene via caesium salts is highly recommended [85].

3.1.3 Analysis of Support-Bound Amino Acid Derivatives

Analytical control forms an essential part of all organic chemical investigations. This is without any doubt not a simple task, however, in synthetic experiments on insoluble gel phases.

As already discussed in the beginning, the chemical behavior of functional polymers is characterized by the heterogeneous and variable microenvironment of each reactive group on the macromolecular thread, which is randomly coiled and fixed in space by cross-linkages. Only with limitations does this situation allow one to apply the knowledge of analytical chemistry in solution to reactions in the gel state. A function on an insoluble but swollen polymer shows not a specific reactivity, but a variable one, depending on the location of the reactive group and its state of solvation. In general, therefore, analytical reactions on polymer are combined with diminished accuracy, and two or three counter-tests with different methods are recommended.

To form a basis for the calculations of excess reagents to be used in a Merrifield synthesis and of reaction rates on the various stages, it is necessary to know the initial load of the gel phase with the starting amino acid or peptide attached onto it, as described in the preceding section.

The initial load can be determined, for example, by

a) balancing the gain in weight,
b) elemental analysis of the nitrogen content,
c) measurement of the amount of remaining binding sites not transformed in the loading reaction,
d) cleavage of the polymer-bound amino acid derivative from a weight sample under the detachment conditions, which will be used at the end of the planned synthesis,
e) cleavage of the N-terminal protecting group of the polymer-bound amino acid or peptide followed by measuring the content of primary amino functions liberated on the gel phase with a specific analytical reaction,
f) measurement of the amount of protecting group fission product cleaved from the polymer.

Difficulties or shortcomings, which will be discussed in the following are inherent in most of these tests. To gain more detailed insight to analytical problems of reactions on insoluble gelatinous phases, some aspects are given below according to

a) The determination of the gain in weight allows only a rough estimation of the loading capacity of a cross-linked polymer, since it is difficult to obtain a really dry gel, absolutely clean and free from any released reaction product, In addition, in some cases the transformation of an anchor group – i.e., chloromethyl- or bromoacetyl functions – is combined with the release of a heavy portion of it (e.g., chloride or bromide), which decreases drastically not only the total gain in weight of the polymer but also the sensitivity of the test.

b) The elemental analysis of the nitrogen content after redoubled combustion of the polymer sample is a reliable measurement of the load as long as a contamination by any nitrogen-containing reaction partner or solvent can be excluded and the amount of bound amino acid derivatives ranges with > 0.8 mmole/g of support. These conditions, however, are seldom realized: For example, in most of the early loading experiments tertiary bases capable of reacting with chloromethyl and bromoacetyl functions were present.

This undesirable side effect, which falsifies the nitrogen content of the loaded polymer, to date is eliminated by the use of tetraalkylammonium- and caesium salts of amino acids or peptides to be bound with their carboxylate function to chloromethyl- or bromoacetyl sites with formation of ester linkages. However, even in those experiments, we still detected a superelevated nitrogen value, which was dependent on the widely used solvent dimethylformamide. At higher temperatures this solvent tends to decompose slowly, evolving dimethylamine, which can react irreversibly with functional sites on polymer resulting in an increased nitrogen content of the support that feigns a higher load with amino acid derivatives, if the load test is based on the elemental analysis of nitrogen.

c) Only in cases of chloromethyl and bromoacetyl sites on polymer can the rate of transformation by loading with amino acid or peptide salts be determined very accurately from the remaining halogen content of the support compared to the original one, assuming that the polymer has been carefully washed from released halogen salts and none of the initial anchor functions are hydrolyzed during the attachment reaction. In the author's opinion the best method for this test seems to be to decompose a weight sample of the loaded polymer with pyridine [70] at high temperature in a sealed flask followed by direct titration of the pyridinium halogenide in the reaction mixture with $AgNO_3$ at a silver electrode.

d) A most realistic analysis of support-bound amino acid derivatives can be seen in the detachment test under conditions selected to liberate a peptide at the end of a planned gel phase synthesis from the polymer support. The released material can be weighed out directly, if a sample of about 2—5 g of loaded support is tested, or a quantitative amino acid analysis can be performed from an aliquot of the detached amino acid derivatives (for released peptides after total hydrolysis) to measure their concentration.

With this method of analysis one takes the risk that not all bound molecules are liberated because some may be fixed in areas of the polymer scarcely solvated under the cleavage condition and therefore they are excluded from reaction. Since this effect depends on the polymeric properties of the macromolecular support, it will not be altered at the binding site by an elongated peptide built up under solvation conditions of a gel phase synthesis, which in most cases differ considerably from those for cleavage. Even at the end of a peptide synthesis those areas are poorly solvated during the detachment reaction and therefore are inaccessible. In other words, this method of load test does not always measure the total capacity of a support, but yields the amount which surely can be cleaved from the polymer at the end of a synthesis.

It should be mentioned in this connection that the conditions of peptide total hydrolysis cannot be recommended to analyze quantitatively the load of a polymer support by amino acid derivatives. Depending on the natur of the amino acids and type of protecting groups on their side functions, the amino acids hydrolyzed in the presence of the support interplay with the macromolecular coil in an unmethodical manner and therefore are not recoverable in their accurate proportions. In general, it is very delicate to wash those polar molecules completely out of a lipophilic polymer tracery.

e) The determination of functional sites like primary amino groups on polymer — to measure the yield of attachment by deprotection of an amino acyl moiety bound to the support — by an additional chemical reaction for analysis is a widely used but not always critically judged testing method. Doubtless, the titration as described on the preceding page of a hydrochloride formed with amino functions [91] is an accurate measure as long as no halogen-containing solvents were used in preceding washings of the polymer support. However, it is not always possible to use this technique to determine deprotected amino functions in the presence of other side groups and remaining binding sites on polymer. In those cases in several laboratories the ninhydrine reaction or Schiff base formation with several aldehydes and the test with picric acid are utilized [92], but all these analytical methods well known to yield quantitative results in solution cannot be applied with the same accuracy on the gel phase (Fig. 31). Since, however, these problems are mainly encountered with the monitoring of deprotection reactions on gel phase syntheses, we shall discuss them in more detail in Sect. 3.2.4.

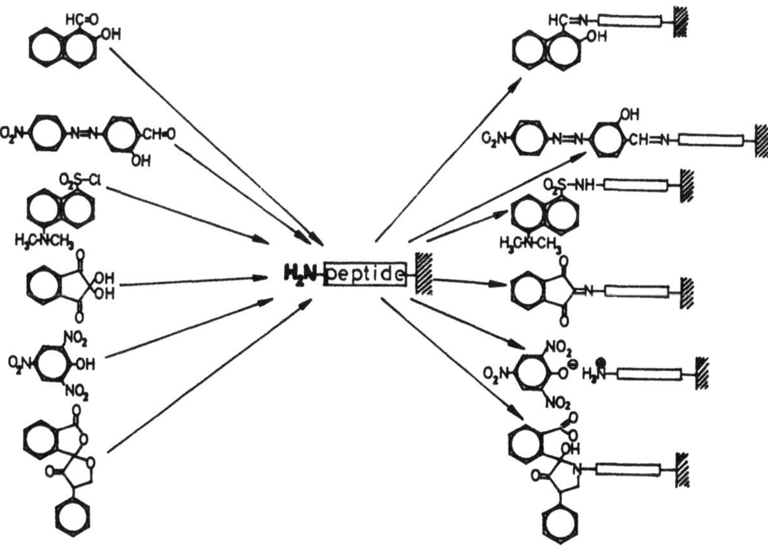

Fig. 31. Analytical color reactions to determine qualitatively the amount of amino functions in polymer phase

f) In our hands a very simple, accurate, quick, and nondestructive test on the amount of amino acids or peptides bound to a support can be performed with the whole batch of polymer on-line in a starting gel phase synthesis, using the amino functions protecting group α,α-dimethyl-3,5-dimethoxybenzyloxycarbonyl (Ddz) [93, 94] (Fig. 32).

These Ddz-protected amino acid derivatives as well as the protecting group fission products show a very specific UV spectrum with absorption maxima at 224—229 nm ($\epsilon = 7,220$) and at 276 and 280 nm ($\epsilon = 2,320$) outside the normal aromatic region not altered by any amino acyl residue [95] (Fig. 33). To determine the load on polymer, the Ddz group is cleaved and the support is washed until a UV test in the final filtrate attains zero. The deprotection reaction and the washings are repeated to control their completion. All collected filtrates are evaporated and the concentration of the Ddz group cleaved is quantified spectrophotometrically. The general utilization of this very helpful principle of detection in the various stages of a Merrifield synthesis will be discussed further in several sections of this book.

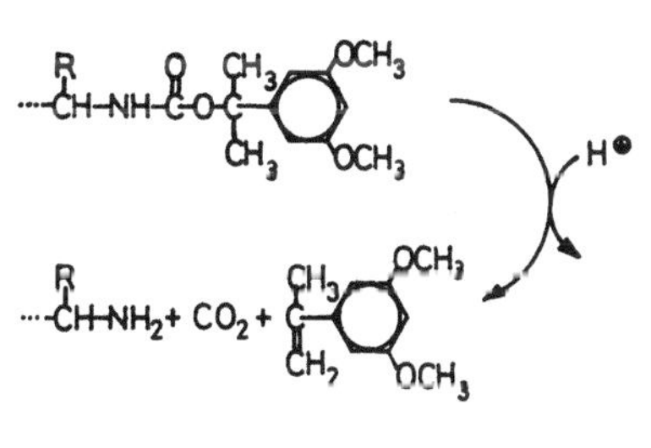

Fig. 32. The α,α-dimethyl-3,5-dimethoxybenzyloxy-carbonyl-(Ddz-) protecting group and its fission product 3,5-dimethoxy-α-methylstyrene

35

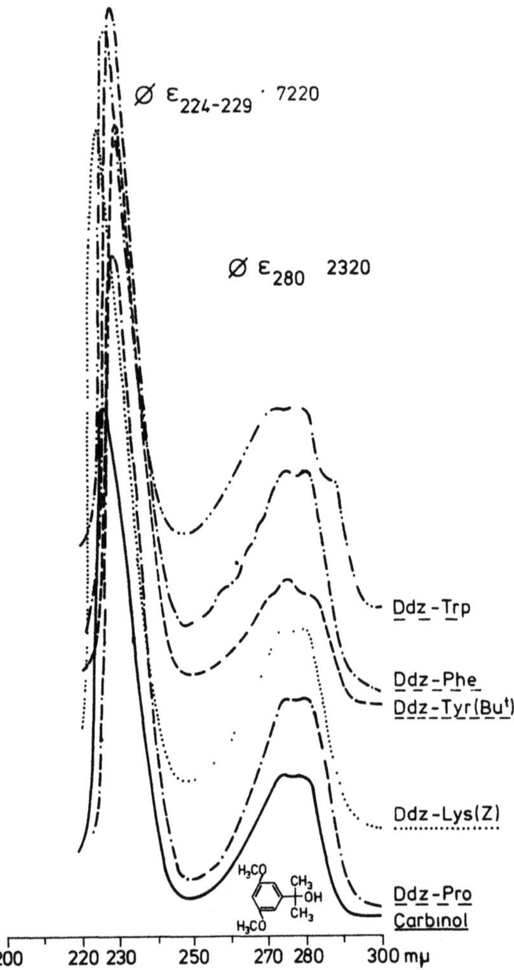

Fig. 33. UV spectra of several Ddz-amino acids

3.2 The Deprotection and Deprotonation

The aspects discussed in the preceding section elucidated the part of work to be finished in batches before starting — more or less automatically — a Merrifield peptide synthesis with its returning cycles of operations on each stage of the process.

The polymer support, loaded with a well-known amount of the first C-terminal amino acid, has to be preswollen and washed with an appropriate solvent, e.g., dichloromethane, before the repetitive procedure of the Merrifield synthesis is initiated with the cleavage of the temporary N-terminal protecting group to liberate the first amino function. Deprotecting reagents and fission products are washed out with an inert solvent. The completion of these steps has to be monitored by a suitable procedure [92, 95] to prevent undesired side reactions. Assuming that the temporary N-terminal protecting group was cleaved by acid, the liberated nucleophilic amino function is still blocked by a proton, which has to be subtracted by admixture of tertiary base. Then the support is rinsed again, as in the beginning, with a highly solvating inert solvent (dichloromethane?) from excess reagent. The amount of amino functions liberated on polymer then should be determined quantitatively before striding ahead to the actual peptide synthesis reaction.

3.2.1 Aspects on Choice of Protecting Groups

The success of a Merrifield synthesis is very strongly influenced by the selection of protecting groups to temporarily shield the terminal amino function and to permanently block side groups of trifunctional amino acids from undesired reactions. Even on polymeric phase, the strategy of a peptide synthesis is determined by the choice of the protecting groups.

We have to differentiate three types of bond strengths of protecting groups (Fig. 34). The C-terminal link has to safely anchor the growing peptide to the polymer support during all chemical operations of the synthesis. Protecting groups on side functions have to be more stable than the C-terminal link. They have to survive under conditions for cleavage of the synthesized peptide from the support, if a completely protected peptide suitable for further fragment condensations is desired. The quality of the temporary protecting groups, however, should allow us to liberate the N-terminal amino function, without harming the C-terminal peptide bond to the support and the masking of side groups.

Predominantly the tert.butyloxycarbonyl (Boc) group [96], in nearly all syntheses published to date, is used to protect temporarily the N-terminus of a peptide to be built up on polymer phase. From recent investigations it is well known, however, that the lability of the Boc groups is not always pronounced enough [208]. Both the C-terminal bond to the support and the maskings on side functions of a peptide are affected to a certain extent if in these positions benzyl ester type of linkages or benzyloxycarbonyl groups are utilized [97]. In those cases, one tries to circumvent this problem of instability by introduction of halogen [98] or nitro [99] substituted benzyl moieties, which because of the electron withdrawing effect of the substituents are more stable under the often repeated acidic Boc cleavage conditions. Yet those stabilized bonds are more reactive to nucleophiles, an aspect which interferes with the chosen strategy from an opposite point of view. Ammono- and aminolytical side reactions as well as transesterifications are possible on benzyl ester groupings. It was therefore very promising to search for more labilized mobile

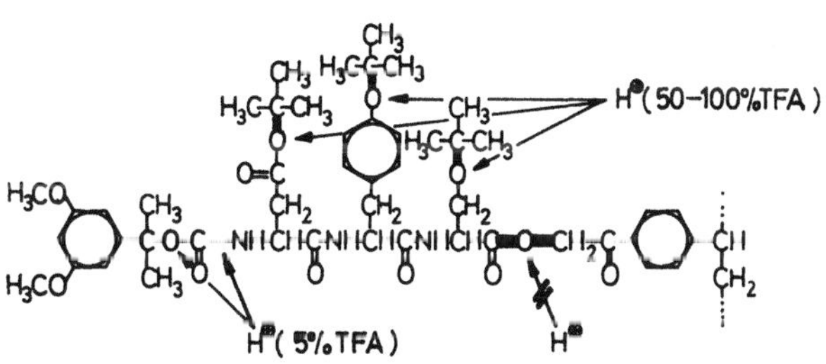

Fig. 34. Three types of bond strengths in protected peptides linked to a polymer phase

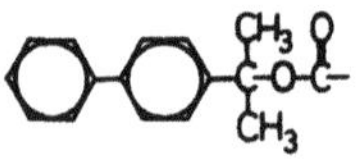

Fig. 35. The biphenyl-isopropyloxycarbonyl-(Bpoc-) protecting group

N-terminal protecting groups than Boc, a concept which would make is possibil to use the lipophilic and sterically shielding tert. butyl principle [100] for masking of all types of peptide side functions instead.

For this reason the biphenyl-isopropyloxycarbonyl (Bpoc) group [101], which can be cleaved completely within 15 min by 0.5% trifluoroacetic acid in dichloromethane (Fig. 35), was developed and recommended. Yet in the meantime, in several laboratories this protecting moiety was found too labile to be of practical use in extended gel phase syntheses, since the acidity of Bpoc-amino acids themselves is sufficient to deprotect the molecule autoacidolytically.

For the same purpose of selective N-terminal deprotection in the presence of uninfluenced tert. butyl type masking functions on side groups of the peptide, we synthesized the α,α-dimethyl-3,5-dimethoxybenzyloxycarbonyl (Ddz) group [93], which is at least twice as stable as the Bpoc moiety (Fig. 36). The Ddz group can be cleaved completely from peptides bound to solvated gel phases within 15 minutes by 5% trifluoroacetic acid in dichloromethane; in solution the deprotection is completed within 8 minutes. Thus Ddz fulfills the demand for selective cleavage without harming tert. butyl protected side functions in the form of esters and ethers. Benzyl ester type of linkages, e.g., to the polymer support, are completely stable under Ddz fission condition. Ddz-amino acids are stable against autoacidolysis in solid form and in solution as well. To prepare the protected amino acids, Ddz-prestages like -hydrazide and -azide, both crystalline and absolutely nonexplosive, are commercially available [102]. In addition, a number of suitably protected trifunctional Ddz-amino acids are produced comparable in price to the Boc-derivatives formerly used [102].

Not only because of its moderate lability, the Ddz-group is particularly useful in the practice of the Merrifield synthesis. As only briefly indicated on p. 35, the spectroscopical properties of this masking residue allow the continuous control of the whole Merrifield procedure by flow photometry in addition to quantitative load measurements and detections of completed reactions on gel phase [95, 103]. (For further discussion see p. 45).

In this connection another utility of the Ddz-group should be mentioned briefly. In dioxane and tetrahydrofurane solutions of Ddz-amino acids and peptides the masking group can be cleaved continuously and quantitatively in a flow procedure by photochemically induced autoheterolysis with light irradiation [93]. This possibility of deprotection is particularly useful in peptide syntheses with Ddz-amino acid active esters insolubilized on polymer phase [104]: By aminolysis the Ddz-amino acid is transferred from its support onto an amino component in solution. The dissolved Ddz-peptide so formed can be depro-

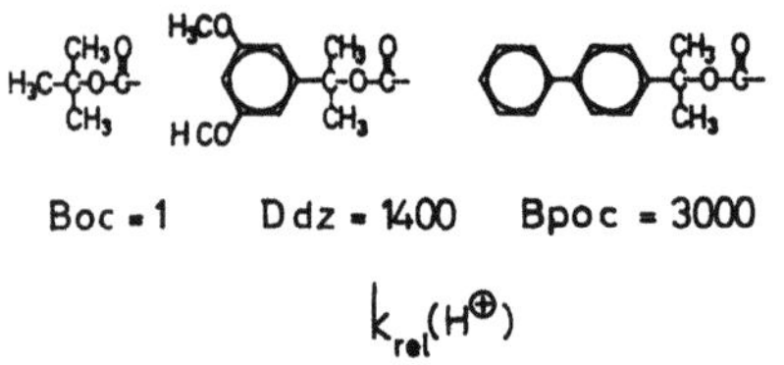

Boc = 1   Ddz = 1400   Bpoc = 3000

$k_{rel}(H^{\oplus})$

Fig. 36. The relative acid stability of the Ddz-protecting group

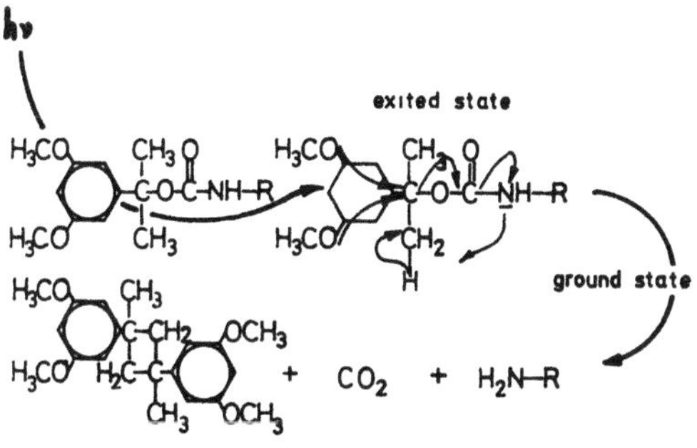

Fig. 37. The photochemically induced Ddz-deprotection for the purpose of peptide synthesis with polymer-supported Ddz-amino acid active esters

tected continuously in the photolytical manner described above to generate a new amino component, which is elongated by reaction with another gel phase-bound Ddz-amino acid active ester yielding the Ddz-peptide with one more residue at the N-terminus of the sequence and so forth (Fig. 37). Since this concept of peptide synthesis utilizing a polymer support as donor of the carboxylic component is opposite to the Merrifield principles, we shall not discuss it here in more detail. This example was given merely to demonstrate the advantageous properties of the Ddz protecting group for different synthetic purposes.

To summarize, the masking functions in a Merrifield synthesis should be selected in such a gradation of lability that the often repeated N-terminal deprotections endanger neither the C-terminal link to the support nor the masked side functions of the growing peptide. The type of anchoring used should allow us to liberate the peptide in an entirely protected form from its support, to allow further use of the sequence in fragment condensation reactions.

These requirements are very well realized by the following choice of protecting groups:
1. Ddz for temporary blocking of the N-terminus, combined with the possibility to continuously monitor all operations during a Merrifield synthesis (see p. 45—46 and 78).
2. Tert. butyl masking to yield ester and ether type side functions. The tosyl group is recommended to protect the guanido moiety of arginine [105]; the benzyloxycarbonyl (Z) group should block the ϵ-amino function of lysine [106]. Thiols are masked by acetamidomethyl-(Acm) [107], tert. butylmercaptan [108], and p-methoxybenzyl groups [109, 209]. Amides are protected by benzhydryl moieties [110], since the 4,4'-dimethoxybenzhydryl [111] was found slightly labile under Ddz-cleavage conditions [112] (the same is the case for trityl masked thiols).
3. To this selection of protecting groups the benzyl and phenacyl ester type of C-terminal anchoring of the peptide to the support is best fitting. These linkages are entirely stable under acidic Ddz cleavage conditions and the latter renders the detachment of fully protected peptides possible in fairly good yield, by catalyzed transesterification and saponification, an aspect which will be discussed in more detail in Sect. 3.5.

3.2.2 Deprotecting Reagents

There are very few examples in which masked N-termini of polymer-supported peptides for practical purposes are deprotected by exchange reactions like thiolysis [113] of nitrosulfenyl groups and hydrolysis [114] of trifluoroacetyl residues.

Since the majority of N-terminal protecting groups are urethanes which are labile to acids, a further selectivity in the liberation of amino functions can be exploited, varying the acidity of the deprotecting reagent. Principally all acid solutions employed need the best solvation properties for the peptide-on-polymer, to prevent incomplete reactions of the protected sites. For this purpose, solutions of the distinct lipophilic and polar trifluoroacetic acid (TFA) in dichloromethane or chloroform are very well suited for deprotection reactions on polystyrene supports. Depending on the type of masking functions to be cleaved, only the concentrations are varied, e.g., Bpoc: 0.5%, Ddz: 5%- and Boc: 50%-TFA in dichloromethane. In 2% cross-linked polystyrene, (which is more rigid than supports mainly used to date like those with a content of 1% or − as in the author's group − 0.5% divinylbenzene), one needs more polar reagents to solvate at least the bound peptide. Acidic mixtures like solutions of gaseous hydrogen chloride in acetic acid (1N) or in dioxane (4N) are used under those circumstances in which the diminished ability of the dense macromolecule itself to become solvated is taken into consideration. A more polar support, like the recently described cross-linked N,N-dimethyl-polyacrylamide in these hydrophilic deprotecting reagents along with a bound peptide, swells completely and therefore all parts of the polymer are nicely accessible and reactive in the polar deprotecting medium.

In general one has to realize that each deprotection reaction is specifically influenced by the chain length of the polymer-bound peptide and by the acutal N-terminal amino acid residue, because of the differences in solvation of the individual amino acids and of the whole peptide in relation to the properties of the support. For this reason, for example, 5.35 N HCl in dimethylsulfoxide/dichloromethane (1:1) was employed [115] to deal with the problem of different solvations. In another example, mercaptoethane sulfonic acid [116] was utilized in diluted solution to deprotect N-terminal Boc-tryptophane without harming the indole moiety.

These aspects clearly indicate that deprotections on polymer phase never should be considered standardized reactions which can be simply programmed in a uniform manner [117]. From the author's point of view we can only meet this problem by continuous recording of each deprotection reaction, to monitor its completion and to gather additional information on the solvation behavior of the growing peptide on polymer. As shown in the paragraph after next, this is realized in the use of α,α-dimethyl-3,5-dimethoxybenzyloxycarbonyl (Ddz) protected amino acids on polymer phase with the aid of a centrifugal reactor, which will be explained in Sect. 4.1.3.

Series of reagents are in use to deprotect specifically peptide side functions following a Merrifield synthesis. Since this aspect, however, is not a problem in connection with the peptide synthesis on polymer phase, the discussion is omitted here. Questions relating to cleavage of protected peptides from the polymer support are discussed in Sect. 3.5.

3.2.3 Deprotonation Problems

The N-terminal amino functions, after deprotection with acidic media, are still blocked by protons which have to be subtracted before the next nucleophilic reaction of the liberated amino group with another amino acid derivative for chain elongation is possible. This is usually done adding tertiary bases, and in most cases triethylamine is employed for this purpose. Since from conventional peptide syntheses in solution the risk of racemization catalyzed by triethylamine [118] is well known and is circumvented using sterically hindered diisopropyl-ethylamine [119] or N-methyl morpholine [120], some research groups which even in gel phase syntheses are utilizing these very expensive tertiary bases [77, 121]. However, in N-terminal stepwise peptide chain elongations on polymer support, no contact should exist between the activated carboxylic component and the tert. base during the peptide bond formation, since the deprotonation reaction followed by intense washings is separated as a preceding step from the peptide synthesis reaction itself. Therefore a racemization favoured by an oxazolinone [122] mechanism initiated by strong bases like triethylamine is not possible and there are not even any indications for direct C^α-proton subtraction during N-terminal deprotonation by triethylamine.

A more pronounced risk is to be seen in base catalyzed transesterifications and trans-amidations on unmasked polymer-supported peptide side functions or by traces of alcoholic solvents, for example, remaining in the gel from washings preceding the deprotonation. Autoaminolysis of the C-terminal ester bond which links the peptide to its support followed by release of diketopiperazines was detected during deprotonation on the dipeptide stage, mainly depending, however, on conformational singularities of imino acids like proline and sarcosine peptide bonds which favour the ring closure reaction [123—128].

Two more hidden risks during deprotonation which seem to be ignored in most of the Merrifield syntheses published, have to be elucidated. The deprotonation reaction on polymer proceeds with solutions of tertiary bases in solvents like dichloromethane, chloroform, or dimethylformamide. The effect of this reaction is the liberation of very nucleophilic and basic centres in the support-bound peptide, namely, primary or secondary amino groups. All the solvent mentioned above, however, are nothing but masked carbonyl compounds which slowly tend to decompose influenced by light, moisture, temperature, and strength of bases present, forming formaldehyde, phosgene, and formic acid (Fig. 38). Yet these highly reactive products of decomposition, small in molecular size, can irreversibly block the very nucleophilic amino functions just liberated on polymer by formation of Schiff bases. In our group, like others, we detected some indications for this very unfavorable side reaction in several syntheses like decrease in yield combined with faint or more pronounced orange coloration of the polymer. From this point of view the solvents mentioned on this particular stage cannot be considered inert as is generally demanded for all solvents in the Merrifield procedure. Therefore in these cases the recommendation of Sheppard and co-workers [65] should be taken into consideration, namely to use dimethylacetamide as solvent in the deprotonation reaction, which is much less labile than dimethylformamide.

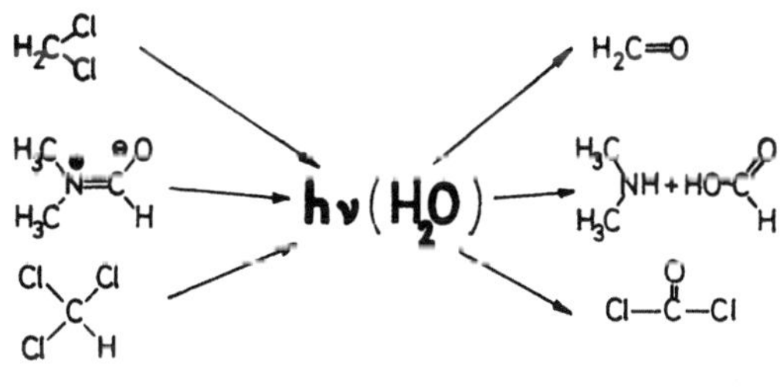

Fig. 38. Several solvents used in the Merrifield process are masked carbonyl compounds suitable to irreversibly block polymer bound amino functions

Finally we have to look on the salt released during deprotonation which usually is triethyl ammonium trifluoroacetate or other combinations of ions corresponding to the acid used for deprotection and to the base employed in the deprotonation step. In general, excessive reagents and reaction products in the deprotonation step are washed out of the support by three or five times repeated agitations with pure solvent like dichloromethane. However, if one records the IR spectrum of the washing media continuously with the aid of a flow cell, one can detect certain amounts of salt released from polymer [103] even after ten rinses. If only a few washings are programmed after deprotonation in the subsequent peptide synthesis reaction, not only can the next amino acid added as dissolved N-protected carboxylic component be activated. By any condensing agent admixed simultaneously also remaining amounts of anions from incomplete rinses after deprotonation are subject to activation. If, for example, trifluoroacetate is present in the activation mixture to a certain extent, free amino functions on polymer may be blocked by trifluoroacetylation, resulting in decreasing yield of the peptide synthesis. Certainly this is the case with remaining acetate ions causing acetylation (Because of this danger see p. 43—44 and 79). Therefore we expressly recommend to control even the completed wash-out of salts formed during deprotonation by IR tests of final filtrates, before proceeding to the actual peptide synthesis reaction.

3.2.4 Analysis of Deprotection and Deprotonation

The homogeneous growth of a peptide during synthesis on polymer phase is determined by the completion of all of the chemical reactions repeated alternately on each stage of the synthesis, namely liberation of the N-terminal amino functions and their condensation with the next amino acid to elongate the sequence. Therefore the control for completion of these reactions is the principal demand in a carefully performed Merrifield synthesis.

Yet we have to differentiate between the two tasks to determine a maximal amount of amino functions on polymer to monitor the total capacity of the support, e.g., after deprotection/deprotonation on one side, and on the other to detect only traces of amino groups remaining eventually after the peptide coupling reaction. In the first case the analytical methods employed are not limited by the presence of labile N-terminal protecting groups as in controlling the completion of the peptide synthesis step.

As already indicated on p. 33 there are several proposals to detect quantitatively the amount of liberated amino functions [92]. However, all of them have to be discussed from the point of view of practical utilization on all stages of a Merrifield synthesis, and in this paragraph we are mainly concentrating on the aspects of analysis of deprotection.

It seems paradoxical to try to monitor quantitatively reactions on polymer phase — which are eventually incomplete because of the heterogeneous locations of functional sites — with additional chemical reactions of similar uncertainty. This fundamental contradiction cannot be abolished by any chemical method of test employed on insoluble gel phases generating a colored reaction on polymer (see Fig. 31). Here, with an additional chemical reaction, a specific dye has to be released and measured as is the case in the quantified [115] ninhydrin [129] test and in the Schiff base formation with 2-hydroxynaphthaldehyde [130]. Although both methods are discontinuous, time consuming, and destructive, they are nevertheless suited to monitor qualitatively the liberation and acylation of amino functions on polymer in specific examples, but not in general. With peptide N-termini, ninhydrine does not always form the Ruhemann purple dye, but stops on an intermediate state of the color reaction. In addition, the initial Schiff base formations of both reagents mentioned above are known to be equilibrium reactions (for further information the brief survey of Beyerman and co-workers [92] is recommended). Radioactive labelled reagents to determine amino groups quantitatively are even less valuable because of the low accuracy of short time liquid scintillation counting measurements.

Ionic reactions [91, 131], if anything, meet the required accuracy to determine a total amount of liberated amino functions on polymer much better than any covalently reacting analytical chemical. Salt formation of amino groups by protonation with subsequent titration of the anions released after deprotonation is almost the best chemical test to determine quantitatively the total amount of amino functions on polymer support, as long as, first, no other basic centers are present which can become protonated and, secondly, no by-products from prestages of the gel phase synthesis contain or release anions of the type on which the quantitative titration is based.

The test in its modified form is performed with pyridine hydrochloride which converts the liberated amino groups on polymer again into the hydrochloride. After extensive washings the salt is released from its anchor functions by deprotonation with triethylamine, followed by repeated rinses of the polymer. The filtrates of this particular deprotonation and subsequent washings are collected and the chloride concentration therein is measured by potentiometric titration with silver nitrate.

Chloromethylated and bromoacetophenyl polymers with remaining functional sites after binding of the first amino acid, and solvents mainly used like dichloromethane and chloroform have to be considered shortcomings in relation to the mentioned chloride titration technique, since all components may generate halogenide ions by side reactions and decomposition, which would falsify the measurement. This is partly circumvented by utilizing pyridine hydrobromide [132] instead of the chloride to perform the salt exchange with liberated amino groups. The bromide ion can selectively be titrated potentiometrically besides chloride from side reactions.

However, like other methods of test to be discussed in the following section, the salt formation with pyridine hydrohalides in the control of completed amino acylation of amino groups unfortunately interferes in the peptide synthesis with masking functions labile to acid [133] and — as already mentioned — with other basic centres on polymer like guanidyl, indolyl, and imidazolyl residues.

A pronounced case is in the direct potentiometric titration of amino functions on polymer, since the titrant is an about 0.5% solution of 72% aqueous perchloric acid in trichloroethane/ethanol, which is added to the whole batch of polymer in a mixture of glacial acetic acid/dichloromethane (1 : 1) [134]. One titration lasts about 1 hour because of the slow reaction of the glass-calomel combination of electrodes on changing potentials and is usually repeated once. To neutralize free amino groups on polymer, each equivalent of perchloric acid of the titrant transports about 2 equivalents of water into the reaction mixture. These moist conditions endanger extended Merrifield syntheses with trifunctional amino acids by proton catalyzed side reactions including detachment from polymer as well as partial acid hydrolysis of the peptide as we found. Since each titration has to be followed by another deprotonation with tertiary base while all polar groups on polymer may be covered with at least one molecule of water, the peptide additionally is endangered by this extended action of base in the presence of water. This may cause detachment, saponification of side group ester functions, and hydrolysis of residual chloromethyl anchor sites on polymer with the consequences discussed on p. 41. Nevertheless, Brunfeldt and co-workers elaborated this method of test and integrated it into a computer-controlled Merrifield process, in which the tiration signals are fed back to the monitor system to realize an automatic peptide synthesis of, however, to date short and lipophilic sequences almost exclusively have been built up from bifunctional amino acids [126, 134].

The conflicting situation of on the one hand most accurate individual titration results accompanied on the other hand by test conditions, which in their necessary frequent repetition chemically endanger the success of the synthesis on the whole, is somewhat similar in the judgement of the other nondestructive method of analysis on polymer-bound amino functions performed on the whole batch, namely the test by means of picric acid [84, 135, 136]. The very acidic 2,4,6-trinitrophenol (pK 0.38) forms picrates not only with N-terminal amino functions, but other basic centers on the support as already mentioned. Principally even this method of analysis interferes with the strategy to use N-terminal temporary masking groups as labile to acid as biphenylisopropyloxy-carbonyl (Bpoc) and α,α-dimethyl-3,5-dimethoxybenzyloxycarbonyl (Ddz), though the picric acid is applicated as its diisopropylethylammonium salt. This aspect will be discussed in more detail in connection with the monitoring of peptide bond formation. In addition, the reagent in repetitive utilizations was found to discolor the polymer irreversibly because of the distinct electrons accepting character of the nitro-substituted phenol, which may form π-complexes. To determine, for example, the completion of a deprotection/deprotonation, the amount of polymer-bound picrate has to be released by extended rinses with tertiary base in dichloromethane. This has the consequences already discussed in connection with deprotonation problems. In an extreme example after 3 hours of repeated 10-min washings, we were not able to completely liberate the picrate from a 5 g batch of polymer-supported peptide. To sum up, the picric acid test on librated amino functions in selected cases is very useful to monitor the deprotection/deprotonation step of a gel phase synthesis. In general, however, the Merrifield concept becomes confined to a certain extent by the picrate determination as well as by the perchloric acid titration: The utilization of acid labile protecting groups, which characterize a most progressive concept in peptide synthesis, and the presence of basic side functions as well as of base labile bonds are prohibited by these methods of analysis on polymer phase.

Nevertheless, Merrifield and his co-workers introduced the picric acid test into a commercial peptide synthesizer to refine it into an almost fully automatic system controlled by a spectrophotometric unit, which determines the concentration of picrate released from the polymer support [137]. Though the machine performs the physical measurements with the highest accuracy available in this connection to date, the best analytical device cannot compensate inherent variable chemical deviations of a test reaction employed on polymer phase.

Based on the results of our investigations and compared to the aspects discussed above, the author holds the view that the qualitative course and a quantifiable analytical control of reactions on insoluble but swelling supports, which are characterized by the heterogeneous location and varying solvation of functional sites, can be performed solely by direct monitoring of the consumption of dissolved reactants or of the release of reaction products from the polymer into solution during the individual steps of a gel phase synthesis, since these are the only parameters, which directly depend on the succeeding synthetic processes in question, although such determinations are considered indirect.

For this purpose we make use of the spectroscopic properties of the acid labile Ddz-protecting group [93], which is removed with 5% trifluoroacetic acid in dichloromethane within 15 minutes. Ddz-amino acids and the masking group fission products 3,5-dimethoxy-α-methylstyrene and 3,5-dimethoxyphenyldimethylcarbinol show identical specific UV spectra characterized by maxima at 224–229 (ϵ = 7,220) and at 276 and 280 nm (ϵ = 2,320) (see Fig. 33). A UV flow cell of a dual beam photometer is perfused continuously with reaction liquid pumped by the mode of action of the centrifugal reactor (for further description see page 75), which contains the polymer, while the reference flow cell remains filled with the original fluid metered by the peptide synthesizer and distributed to the reactor [95].

During deprotection, the recording (Fig. 39) of the synthesis control system shows the increasing concentration of the protecting group fission products released from polymer. The end of the removal is indicated by the horizontal course of the reaction curve. The cleaved product is washed out of the polymer by repeated rinses with fresh solvent, and all filtrates are collected in a separate flask. This procedure is documented on the recorder by typical wash-out events till the control system attains zero on the recording. To control the completion of the operation, the whole process is repeated. Quantitative removal of the Ddz-group in the first reaction period is indicated during repetition by a constant zero line on the recording. In several cases, because of the shape of the kinetic curve we were able to detect not only extended reaction times, but even incomplete removal of the protective group in the first reaction period which clearly indicated alterations in the solvation properties of the peptide-on-polymer and the necessity for repetition of the process. The quantitative determination of the Ddz-fission product is performed spectroscopically in the filtrates collected during deprotection. The results of those measurements yield directly the amount of amino acid incorporated to the polymer-bound peptide in the preceding synthesis step. Since the accuracy of the determination depends on the spectroscopic measurement and not on the reliability of an additional chemical reaction, the quality of this analysis is better than that of any other method of test.

Our continuously operating synthesis control system indicates qualitatively even the succeeding deprotonation reaction on polymer phase, since the recorded alteration of the optical density of the reaction liquid circulating through the flow cell of the photometer is caused by salt formation during liberation of the amino functions on polymer. A constant level on the recording in repeated washes with tertiary base solutions indicates no further transformation of tertiary base, which means completed liberation of the amino groups on polymer.

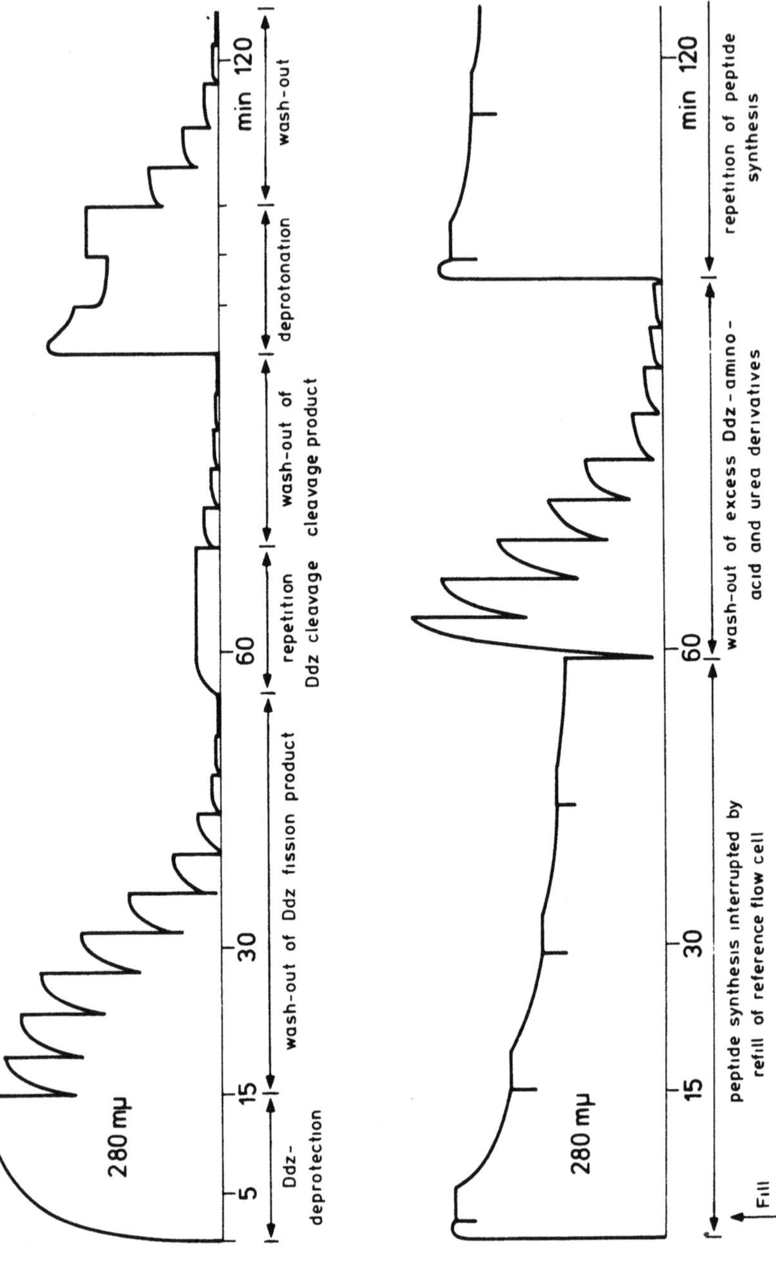

Fig. 39. The recording of the continuously operating photometric synthesis control system shows the course of the proceeding Merrifield synthesis

3.3 The Peptide Bond Formation

After deprotection and liberation of the amino groups, the polymer-bound peptide is prepared to become acylated by another N-protected building block, which usually is an amino acid. The dissolved carboxylic component is activated before the admixture with the polymer or in the presence of it, followed by intense mixing of the phases.

Most of the activation methods known from conventional peptide synthesis in solution were explored on polymer concerning their desired reactivity in relation to side reactions, which are particularly unfavorable if the by-products are also bound to the support. Though urethane-masked amino acids generally are shielded from racemization under controlled conditions, this problem limits the use of peptidic building blocks C-terminally activated, since they tend to form the redoubtable oxazolinone intermediate from which the abstraction of a C^α-proton is facilitated (Fig. 40). Recent results indicated a chance to overcome this problem and will be mentioned in the proper section of this chapter.

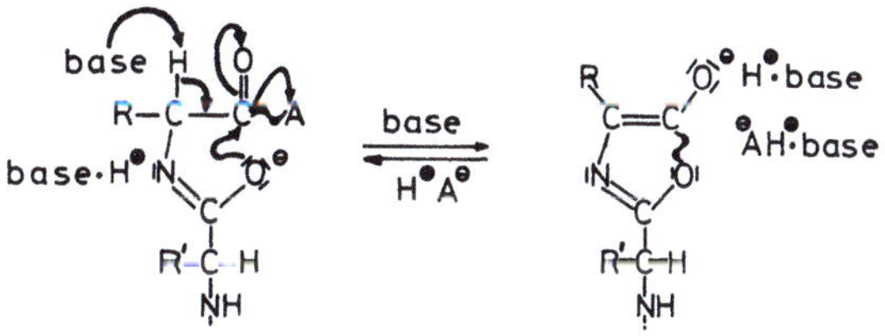

Fig. 40. The oxazolin-5-one mechanism of C-terminal peptide racemization

To meet the fundamental demand for completion of all reactions on polymer, the activated carboxylic component usually is added in great excess, which naturally enhances the possibility of side reactions on the solute and on polymer as well. In the following, the main methods employed in the Merrifield synthesis, including those very opposing aspects like complete transformations, however free from side reactions, have to be judged and taken into consideration.

3.3.1 Dicyclohexylcarbodiimide Activation

From the very beginning to date dicyclohexylcarbodiimide [89] has been by far the most utilized reagent to activate N-protected amino acids, because of its convenience in use and its very suitable reactivity. Usually the carboxylic component and the lipophilic diimide in equimolar portions are admixtured to the polymer-bound amino component in a three to fivefold excess, in most cases dissolved in dichloromethane, which was found not to favor racemization and to suppress other side reactions during activation. To approach, as closely as possible, the desired complete transformation on polymer, the dissolved compounds are introduced in high concentrations so that the volume depends solely on the swelling rate of the gel phase, which has to be well suspended in the reaction liquid.

Under those conditions, the highly reactive O-acyl isourea in a first intermediate stage is formed, initiated with protonation of the diimide by the carboxylic component.

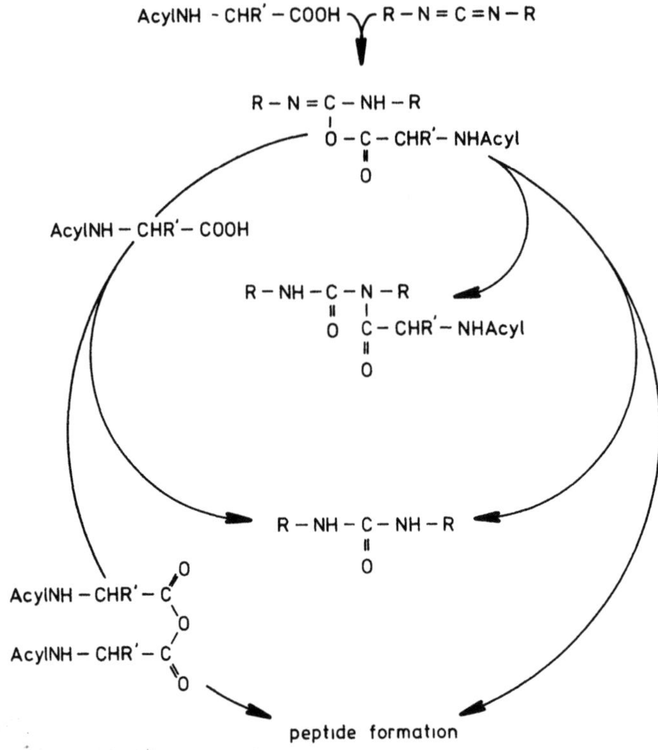

Fig. 41. The branched pathway of reactions with dicyclohexylcarbodiimide and N-protected amino acid carboxylic functions

The amino functions on polymer in a nucleophilic attack react with this acylating intermediate from the moment of its formation. In contrast to the situation in real solution, the reactivity of the amino groups in the gel phase is lowered, leaving time for a branched pathway of side reactions of the O-acyl isourea. It is attacked by another carboxylic partner forming the symmetric anhydride as a second highly reactive intermediate, which acylates the amino groups on polymer. In both the direct and the indirect acylation, dicyclohexylcarbodiimide loses its condensation potency by transformation into the dicyclohexylurea, which in most of the solvents employed in the synthesis step of the Merrifield procedure is only slightly soluble (Fig. 41).

As a further consequence of the high reactivity of the excessive O-acyl isourea in the heterogeneous peptide synthesis mixture, a base-catalyzed intramolecular O-N-acyl migration takes place, forming the inactive N-acyl dicyclohexylurea. At the same time, this by-product is formed by acylation of already formed still dissolved amounts of dicyclohexylurea, which can be attacked by the symmetric anhydride or the O-acyl isourea of the carboxylic component as well. Both the N,O-acyl shift and the latter side reaction decrease the total concentration of the activated masked amino acid; even N,N'-diacyl derivatives of the urea also can be formed as further by-product. Fortunately, all of these acyl urea derivatives are not fixed to the polymer phase and are well soluble in dichloromethane, so that they can be washed out easily from the gel phase after the

48

peptide synthesis. By far, it is not possible with the same simplicity, to rinse out the precipitated amounts of dicyclohexylurea, which sticks even inside the polymer beads in the tracery of the macromolecular coil. The most efficient washing would proceed in the swollen state of the matrix. This would render it possibile to dissolve the interior deposit of the dicyclohexylurea. Unfortunately, the urea is best soluble in methanol that shrinks the polymer. We found a mixed system of dichloromethane/methanol 4:1 (v/v) very suitable for this purpose, since it holds the gel support completely solvated and dissolves about 1 g dicyclohexylurea in 100 ml of the solvent. Naturally the methanol content of the polymer has to be washed out afterwards with pure dichloromethane, to prevent solvolytic side reactions. For this reason, other laboratories admixture, as we formerly did, sterically hindered alcohols like isopropanol, tert. butanol and tert. amylalcohol. However, these liquids all dissolve the dicyclohexylurea less efficiently and do not solvate the polystyrene matrix, though they are more lipophilic than methanol.

In a shortened synthesis program one can combine the wash-out of the urea derivative with the subsequent deprotection reaction [61]. The acidic reagents used on that step, such as trifluoroacetic acid/dichloromethane or hydrogen-chloride/acetic acid, are excellent solvents for the urea by-product. Under these circumstances, however, the repetition of the actual peptide synthesis step, which eventually enhances its completion, is no longer possible, since the next cycle of operations is already initiated with the N-terminal deprotection of the just elongated sequence. We utilized this time-saving, shortened program successfully in the synthesis of several uncomplicated decapeptide precursors of the cyclic antamanide from Boc-amino acids.

A more hidden and polymer-bound side reaction during activation can be envisaged in the direct reaction of the protonated carbodiimide with amino functions on polymer forming guanido groupings, which block the peptide synthesis [138]. This side product also might derive from the occasionally observed cleavage of an urethane protective group from the N-terminus just synthesized. This might be due to the addition of the carbodiimide onto the masked secondary amino function with formation of the guanido grouping and simultancous removal of the protecting group (Fig. 42).

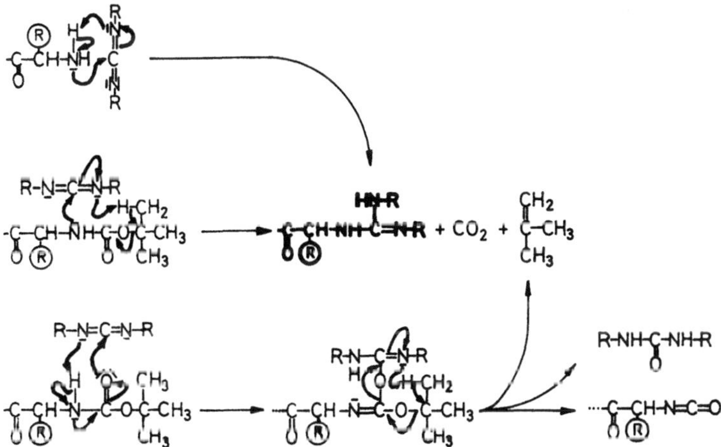

Fig. 42. Carbodiimide side reaction on primary and protected amino functions

The most dangerous contamination of the peptide fixed to the gel phase exists in
the incorporation of the specific amino acid racemized to a small extent by the method
of C-terminal activation. In several investigations it was demonstrated, however, that
dicyclohexylcarbodiimid activation in dichloromethane of urethane-protected amino
acids — with the exception of histidine — beyond room temperature level does not cause
significant racemization [210], which is enhanced, however, in dimethylformamide as
solvent. In cases of apprehended racemization, additives like N-hydroxysuccinimide [43]
or — more recently explored — 1-hydroxy-benzotriazole [44, 139, 140] are in use, which
admixtured in two equivalents per one of the diimide, intermediately form the activated
ester with a considerably diminished tendency to racemize. This technique will be described
in Sect. 3.3.5, since it was shown to be a helpful method to activate even peptide fragments
without racemization.

Also, side reactions from the amide groupings of asparagine and glutamine — that are
caused by initial nitrile formation during dicyclohexylcarbodiimide activation — are pre-
vented by addition of two equivalents of 1-hydroxybenzotriazole. Usually we introduced
these amino acids with masked amide nitrogene utilizing the 4,4'-dimethoxy-benzhydryl
(Mbh) moiety [111]. Since this protection, however, was found to be not entirely stable
against 5% trifluoroacetic acid/dichloromethane during removal of the temporary N-
terminal masking Ddz-function, slightly cleaved amounts of the Mbh-group disturbed
the accuracy of our photometric monitoring of the synthesis, which depends upon the
spectroscopic properties of the Ddz-residue. Therefore to date, we use the diphenyl-
methyl type of amide protection [110] or β-benzyl aspartate or γ-benzyl glutamate, whose
side functions subsequently to the synthesis are converted into amide groupings.

As already mentioned, one of the acylating species formed by activation with dicyclohexylcarbodiimide has been known for a long time to be the symmetric anhydride of
the carboxylic component. Its formation can be favored by admixture of two equivalents of the N-protected amino acid with one of the diimide. Since the major portion
of the dicyclohexylurea is precipitated, if dichloromethane is used as solvent, the by-
product can be filtered off before mixing of the anhydride solution with the polymer
phase [141]. In this way, in the actual peptide synthesis the chance for side reactions is diminished, since only the dissolved part of the urea by-product, small amounts of the O-acyl
lactim, and traces of the diimide initially are still present. Though in most cases the peptides obtained by this procedure obviously are purer than those from the original method
of activation with equivalent amounts of the components, the in situ formation of symmetric anhydrides with dicyclohexylcarbodiimide does not entirely overcome diimide
dependent side reactions and the wash-out problem of the dicyclohexylurea.

3.3.2 Symmetric and Mixed Amino Acid Anhydrides

Although dicyclohexylcarbodiimide activation to date is still the widely used coupling
method on polymer phase, we searched for an alternate to circumvent all the problems
combined with diimide side reactions and the formation of the scarcely soluble dicyclohexylurea. For this purpose we reinvestigated the direct synthesis of symmetric anhydrides
from N-protected amino acids with phosgene [142–145] (Fig. 43).

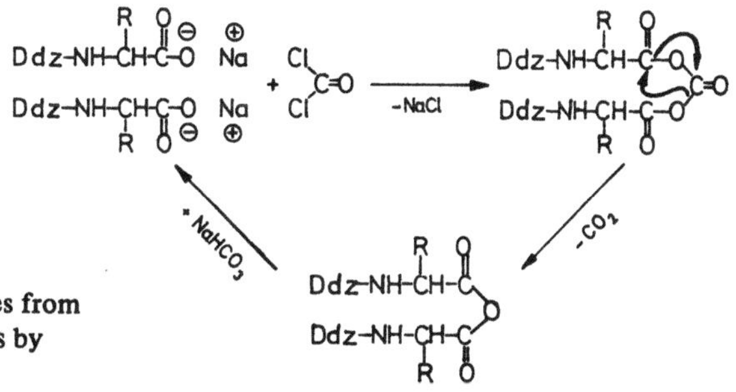

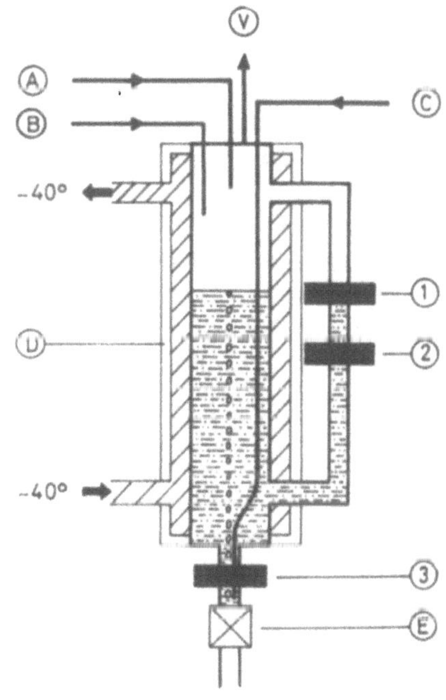

Fig. 43. Synthesis of symmetric anhydrides from
N-protected Ddz- and Boc-amino acid salts by
the aid of phosgene

By the aid of a modified peptide synthesizer solutions of two moles of Boc- or Ddz-amino acid salts (sodium- or triethylammonium form) in tetrahydrofuran or dichloromethane at $-40\,°C$ are mixed for 5 minutes with one mole of phosgene in the same solvent at $-40\,°C$, controlled by the punched tape program of the synthesizer (Fig. 44). The solution of the symmetric anhydride is added at $-40°C$ to the gel polymer, which is contained in the centrifugal reactor (see p. 75) and is allowed to warm up to the cooling water temperature level $(9-13\,°C)$ of the reactor during the peptide synthesis period of 30 minutes. Since the mode of action of the centrifugal reactor renders it possible to perfuse an external IR flow cell with the reaction solution, we are in a position to monitor directly the quality of the symmetric anhydride and its consumption, watching the characteristic absorptions on $\sim 1,830$ and $1,760\ cm^{-1}$ [71]. In several cases the symmetric anhydrides were isolated in crystalline or solid form for characterization (Table 6).

Fig. 44. The modified metering flask of a program-controlled peptide synthesizer for the direct preparation of N-protected amino acid anhydrides with phosgene. (A) Solutions of N-protected amino acid salts in dichloromethane or tetrahydrofurane; (B) Solution of phosgene in tetrahydrofurane; (C) Dry air inlet for stirring; (D) Vacuum jacket; (E) Drain valve; (V) Vent line. Photo sensor levels to meter: (2) The solution of N-protected amino acid salts; (1) The addition of the phosgene solution; (3) The draining of the reaction flask

Table 6. Boc–NH–CH(R)–CO–O–CO–CH(R)–NH–Boc

R	Mp [°C]	Anhydride carbonyl absortions[a]	
–H glycine	70–72	1840 (s)	1760 (m)
–CH$_3$ alanine	95–96	1829 (s)	1756 (m)
–CH(CH$_3$)$_2$ valine	103–104	1820 (s)	1750 (m)
–CH$_2$CH(CH$_3$)$_2$ leucine	70–72	1815 (s)	1750 (m)
–CH(CH$_3$)C$_2$H$_5$ isoleucine	77–79	1820 (s)	1745 (m)
–CH$_2$C$_6$H$_5$ phenylalanine	78–80	1830 (s)	1760 (w)
–CH$_2$SSC(CH$_3$)$_3$ S-tert. butylthio cysteine	86–88	1823 (s)	1743 (m)
–CH$_2$–S–CH$_2$C$_6$H$_5$ S-benzyl cysteine	69–71	1818 (s)	1760 (m)
–(CH$_2$)$_2$CO$_2$CH$_2$C$_6$H$_5$ γ-benzyl glutamate	70–72	1825 (s)	–[b]
–CH$_2$C$_6$H$_4$OCH$_2$C$_6$H$_5$ O-benzyl tyrosine	76–78	1830 (s)	1760 (w)

[a] IR spectral data [cm^{-1}], s = strong, m = medium, w = weak.
[b] Not resolved from ester carbonyl band.

In the use of the symmetric anhydrides prepared in the way described, there is no wash-out problem of by-products, which would cause side reactions like in the dicyclohexylcarbodiimide procedure – except precipitating inert sodium chloride; therefore this method of activation is particularly suitable in the Merrifield synthesis. However, the phosgene method of activation is sensitive: first, against impurities in the starting material, which therefore need analytical grade quality; second, against insufficient cooling; and third, against excess of triethylamine which can be very clearly observed in the IR spectrum.

The most secure procedure for the clean preparation of symmetric anhydrides was found in the use of a 10% excess of free carboxylic component over its triethylammonium salt and a 10% excess of this salt over the equivalents of phosgene added at –40 °C.

Impure IR spectra of symmetric anhydrides (Fig. 45), particularly in the range of 1,690 and 1,800 cm^{-1}, show absorption bands which might be allied to Leuch's anhydrides [146, 147] (excess phosgene) and N-acyl-aziridinones [148] (excess triethylamine). Though the latter was demonstrated to be the highly reactive racemization-free precursor of both the intramolecular asymmetric and the intermolecular symmetric anhydride, the tendency for racemization of the Leuch's compound during reaction with the gel-phase-bound amino component has to be considered. In two cases even a distinct dependence on the salt cation of the carboxylic component in the formation of the possible side products mentioned was observed. Exclusively the sodium salts of Ddz-valine and Ddz-(tert. butyl)-

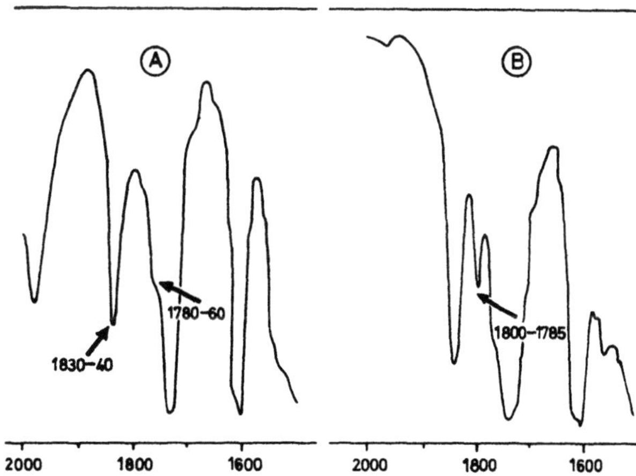

Fig. 45. IR spectra of pure (a) and impure (b) symmetric amino acid anhydrides prepared with phosgene

threonine form pure symmetric anhydrides with phosgene, whereas triethylammonium-, potassium-, lithium-, and tetrabutylammonium salts result in impure IR spectra [71].

Nucleophilic centers of trifunctional carboxylic components naturally have to be masked before the formation of symmetric anhydrides. Ddz-tryptophane, for example, shows several side reactions with phosgene, whereas the same compound with formyl masked indole nitrogen results in an excellently pure symmetric anhydride [71] (Fig. 46).

Under the above mentioned conditions of the formation and peptide reaction of symmetric anhydrides, we have no indications for over-reactions like self-acylation at the urethane-masked amino group of the activated compound [71], as was occasionally observed by Merrifield [149].

Since only one-half of the symmetric anhydride can be transferred onto the polymer-bound peptide by acylation of the amino functions, six to ten moles of the carboxylic component are introduced into the system per mole of reactive sites on polymer. The great

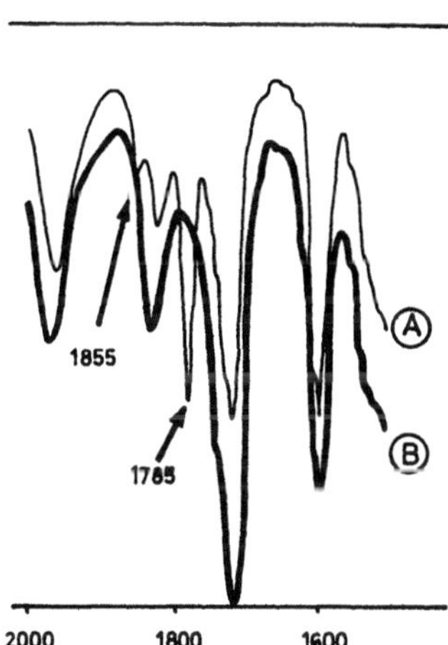

Fig. 46. IR spectra of Ddz-tryptophane anhydride (a, impure) and of Ddz-N^{ind}-formyltryptophane anhydride (b, pure) prepared with phosgene

advantage of this method, however, is the fact that excess N-protected amino acids — which often are very precious — can be recovered analytically pure after bicarbonate hydrolysis from the filtrates of the peptide reaction step on polymer. The average rate of recycling was found better than 80%. Therefore, even multiple repetitions of a specific coupling reaction aiming at complete transformations on polymer are limited neither from an economic point of view nor by the aspect of overloading the polymer with by-products [71].

The coupling method with symmetric anhydrides prepared with phosgene and used in situ with permanent monitoring of the anhydride by flowthrough IR measurement was practized in the synthesis of several linear decapeptide precursors of antamanide. These were obtained in yields between 84% and 96% and in the synthesis of fully protected somatostatin [71], which was built up on polymer with 83% yield as determined by analytically quantified Edman degradation on polymer.

Though the toxicity and vapor pressure of even dilute solutions of phosgene and the necessity for deep-temperature equipment might be considered deficiencies, we have to state that the preparation of symmetric anhydrides with phosgene and their facile and efficient use on polymer phase seem to predestinate this method of activation to the Merrifield synthesis, since all chemical problems depending on dicyclohexylcarbodiimide are circumvented, the wash-out operations (excluding sodium chloride) are simplified, and excess N-protected amino acids are to be recovered pure and in high yield.

The remaining deficiencies mentioned induced us to also reinvestigate the utility of mixed anhydrides in the Merrifield scheme. Like others, we formerly had observed that, independent of the type of mixed anhydride formed with the N-protected amino acid component, even the wrong half of the mixed anhydride causes irreversible blockage of parts of the polymer-bound amino component [150] (Fig. 47).

Contrary to the situation in real solution, the prolonged presence of mixed anhydride near the free amino component, which is diminished in its reactivity because of its polymer-bound nature, enhances the chance for this undesired attack, which results in a stable polymer-bound urethane by-product. From our investigations we had learned, for example, that mixed anhydrides both from isopropyl chloroformate at 0 °C and anisoyl chloride at +10 °C caused on the average 10—20% blocking of the amino component by wrongside acylation on each stage of the peptide synthesis on polymer. This results in a steep decline of the growth of the sequence with total blocking of all amino groups after about ten stages.

If, however, six moles each of 1-hydroxybenzotriazole and N-methylmorpholine are added to one mole of the polymer-bound amino component, prior to the application of

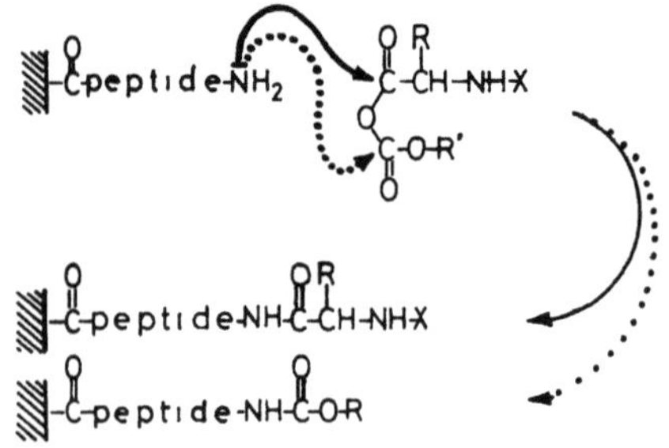

Fig. 47. The reaction pathway of mixed anhydrides of N-protected amino acids in the peptide synthesis in polymer phase

three moles of the mixed anhydride of the N-protected amino acid formed with isobutyl-chloroformate, after one repetition of the coupling reaction which each lasts 30 minutes, there is no blockage resulting from acylation of the wrong half of the mixed anhydride by the aid of a peptide analyzer. On six stages of an LHRH synthesis we measured an average yield of 95%/stage. About 70% of the excess N-protected amino acids can be recovered from the filtrates of each synthesis step [151]. Nevertheless, the general application of this modified coupling method has to be further investigated to prove its versatility.

3.3.3 Active Esters

Because of the side reaction of dicyclohexylcarbodiimide with the amide grouping of asparagine and glutamine during activation, their nitrophenyl esters [152] instead have been widely used for a long time to circumvent this problem, though the slow and incomplete reaction of these active esters on polymer were often stated. During the last decade several research groups looked for a general use of active esters in the Merrifield synthesis, since the action of these esters on polymer might be monitored by the phenolic compartment released in the aminolytic reaction [153], because excesses of the activated N-protected amino acid might be recovered, combined with facilitated wash-out from the polymer, and the active derivatives might be pre-prepared and stored. Nevertheless, at last the addition of catalysts like 1-hydroxy-benzotriazole and 3-hydroxy-4-oxo-3,4-dihydro-chinazoline was found to accelerate the reaction rate of trichloro- and nitrophenyl esters in dimethylformamide on polymer so drastically that their general application became reasonable from the point of view of total reaction times and yields. Even in the modified use of mixed anhydrides mentioned in the preceding section, we detected the formation of an active ester with 1-hydroxybenzotriazole to be the intermediate acylating species. Recently the most reactive pentafluorophenylesters were discovered by Penke to be particularly suited in the Merrifield synthesis because of their acylating efficiency in short time coupling reaction. The versatility was demonstrated in the synthesis of five fragments of ACTH yielding finally 10% of the biologically active hormone sequence 1–24 [154].

3.3.4 Redox Condensation

In the preceding examples of activation methods, the carboxylic component was reacted with a single agent to form the acylating intermediate. Mukaiyama, Ueki, Maruyama, and Matsueda developed an activating system of three components utilizing the reductive potential of triaryl- or trialkyl phosphines on 2,2'-dipyridyl disulfide and the carboxylic component [155]. One equivalent of the thiol, generated during redox reaction with the phosphine, is substituted in the binding sphere of the phosphorus atom by the carboxylate of the N-protected amino acid which thereby becomes reactively bound to the oxidized center (Fig. 48). Because of the tendency of the latter to take over the carboxylate oxygen forming the phosphine oxide with simultaneous release of the second equivalent of 2-mercaptopyridine, the activated acyl residue can be transferred onto an amino component yielding peptide bond formation. All partners in this concerted reaction are very soluble in dichloromethane; there are no wash-out problems of excess materials and the method

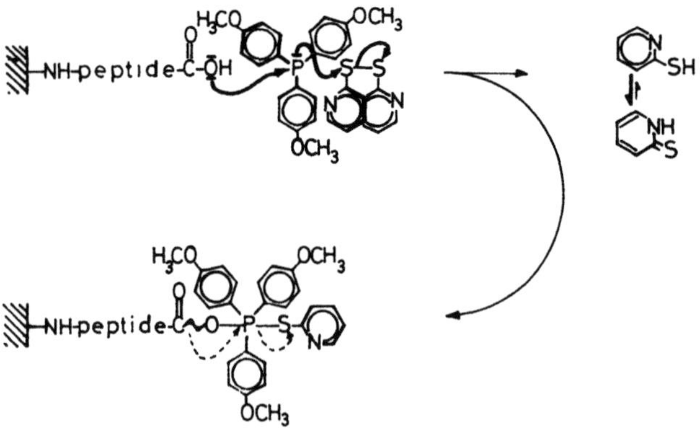

Fig. 48. The C-terminal peptide activation by the redox-components trianisylphosphine and 2,2′-dipyridyl disulfide

is reported to be free from racemization, if urethane-masked amino acids are activated. We found, however, that the recycling of excess material is possible only with great efforts of purification comparable to those combined with the use of dicyclohexylcarbodiimide. Together with Ueki [103] we practized the redox condensation utilizing trianisyl phosphine, which was particularly recommended by him and Ikeda for activating urethane-protected nitro-arginine. With a tenfold excess of the components dissolved in dichloromethane in a 0.3 molar reaction mixture with repetition of the coupling process, which lasted 1 hour each at 30 °C, we synthesized the secretin sequences 18—27 and 12—17 containing four arginine residues, from Ddz-amino acids. Within the tolerance of the photometric determination and by reference to the continual monitoring of the completion of all synthetic operations for the former preparation, quantitative peptide couplings were established. The latter sequence was built up with similar success. Though this method of coupling is reported not to need masked side functions of trifuntional amino acids like serine and glutamine, we did not renounce the solubilizing effect of shielding groups like tert. butyl esters, -ethers and the 4,4′-dimethoxybenzhydryl residue as well in dichloromethane and dimethylformamide. By continuous UV and IR flow-through spectrophotometry, we were also in the position to observe the tautomerism of the 2-mercaptopyridine into 2-pyridine-thione released during peptide synthesis, which is a direct indication of the proceeding redoxreaction at the phosphorus atom. Although in our investigation we did not have to incorporate cysteine into the sequence, the mentioned formation of 2-mercaptopyridine is reported not to require snatching by an additive, since the tautomeric equilibrium is vastly shifted to the thione side as we confirmed by photometric control. The potential of the redox condensation method to date is not entirely explored. This type of carboxylate activation does not favor the formation of an intermediate oxazolinone at the C-terminus of a peptide which would cause racemization. Complete suppression of this side reaction might be reached by further variation of the substituents on the phosphine component. Therefore the redox reaction can be used to couple amino acid tert. butyl esters onto peptides which are bound by their amino functions to a polymeric support, so that the free carboxyl end has to be activated [156]. This way of peptide synthesis, in the opposite direction, was formerly practically forbidden, because of the racemization problem induced by other methods of C-terminal peptide activation. However, it would be of great advantage to utilize the urethane-type of linkage between a peptide N-terminus and the support, since it renders possible a clean and smooth acidolytic cleavage of the synthetic end product from polymer.

3.3.5 Fragment Condensation on Polymer Phase

The aspects discussed in the preceding paragraph raise the question whether building blocks like peptides can be attached to each other on polymer support.

As already mentioned in the beginning, from the statistical point of view it is a great advantage on gel phase reactions if peptides could be used instead of amino acids to elongate a sequence on polymer. In cases of incomplete peptide reactions the following effects would happen. First, the excessively introduced dissolved peptidic component could be easily recovered from the filtrates of the gel phase reaction and, second, the by-product distribution on polymer — as explained in Sect. 2.2 — would be drastically diminished by the small number of transformations. Third — and this is most essential — all by-products on polymer among one another and to the target sequence could differ much more pronouncedly in molecular weight than is the case with amino acid units as building blocks. This would allow a clean separation of the desired polypeptide from shorter sequences, which from a biochemical point of view may possess very similar behavior and activities.

Compared to the situation in real solution, the fragment condensation of peptides on gel phase is handicapped by the polymer-bound nature of one of the components, which hereby thermodynamically becomes less reactive. This effect, however, is by far over compensated by the advantages mentioned above. Yet it has to be stressed that the gelatinous reaction partner for maximal reactivity needs complete solvation of the peptide and its support as well. The term "solvation" means nothing but the gel phase-analogue to the dissolved state of substances in real solution. In most cases, complete solvation of the peptide/polymer conjugate can be realized by swelling it in mixed solvent systems of dichloromethane with dimethylformamide, dimethylacetamide, or hexamethylphosphorictriamide. In addition to the solubility problem, the synthesis of polypeptides by condensation of fragments is generally limited by racemization on the C-terminus of the carboxylic component. This aspect principally is not altered in performing the condensation on gel phase. However, slow reaction rates during peptide bond formation favor this type of conversion. Therefore, the dissolved components have to be added in great excess to the polymeric partner to suppress at least the dependence of the condensation reaction rate upon slow diffusion, which is the other detrimental factor in gel phase transformations with big molecules.

There are only a few methods of C-terminal peptide activation, which during fragment condensation in solution cause on the average less than 5% racemization. Since this danger on polymer phase is enhanced, methods like azide coupling and dicyclohexylcarbodiimide/N-hydroxysuccinimide activation were scarcely applied on gel phase. At last, when the homogeneous catalytic efficiency of 1-hydroxybenzotriazole in combination with dicyclohexylcarbodiimide activation was discovered, an increasing number of fragment condensations on polymer phase were investigated [157, 211]. It was demonstrated that excessively added 1-hydroxybenzotriazole takes over the preactivated carboxylic component forming

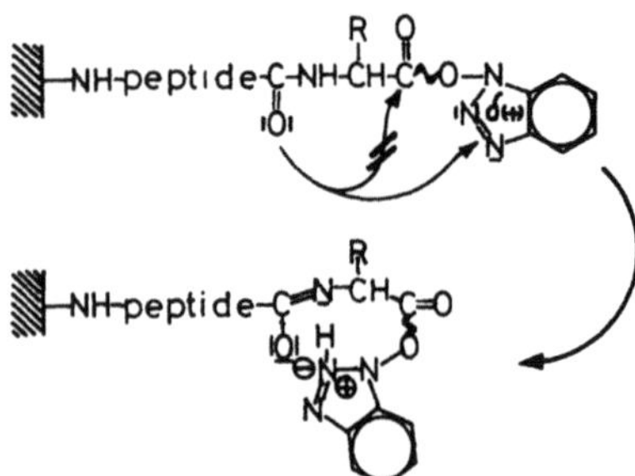

Fig. 49. The racemization suppressing effect of 1-hydroxybenzotriazol in C-terminal peptide activation

an active ester, which — because of the molecular structure of the electron-withdrawing moiety — does not favor racemization even in extended reaction periods on polymeric phase (Fig. 49). The catalytic effect of 1-hydroxybenzotriazole was found highly dependent on polar solvents like dimethylformamide, which is particularly useful both to dissolve peptide sequences and to solvate polymer-bound fragments [44, 158].

The same advantageous properties are described concerning the redox condensation [159]. With this method, the coupling of fragments onto polymer-bound amino- and carboxylic components as well was demonstrated to proceed in high yield without striking racemization. Solvents like dichloromethane and dimethylformamide are both very well suited. The possibility of performing the fragment condensation on polymer at elevated temperature betters the chance for completed coupling reactions. The recovery of the excessively introduced dissolved peptide fragment from the filtrates of the gel phase is facilitated, since the redox products triaryphosphine and its oxide, as well as 2,2'-dipyridyldisulfide and its thione, are easily separated by extraction with ether.

Though in the meantime some successful fragment condensation on polymer are described, we should not overlook the general problem of solubility. A coupling on polymer might have produced in high yield a polypeptide which, however, might be found to be completely insoluble after cleavage of the anchor bond to the support. Such a result has to be considered a dead end of the synthesis. Stepwise fragment condensations should be planned to finally yield the target polypeptide on polymer and not another partial sequence of it, which for further elongation has to be released, for example, from its support entirely masked on side functions. The best chance for solubility exists in the completely unprotected form of a polypeptide which, however, in that form does not allow simple further sequence elongations.

3.3.6 Analysis of Peptide Bond Formation

To date, there does not exist a direct and accurate but nondestructive method to determine quantitatively the formation of peptide bonds on polymer. Without regard to the principal discussion of the problems of chemical analyses on heterogeneous phase (see p. 42), the detection of remaining uncoupled amino groups on polymer does not necessarily allow one to deduce the yield of support-bound peptide. Additional side reactions might have blocked a certain amount of functions; fission reactions might have cleaved parts of the peptide, or new functional sites might have been introduced, which positively interfere with the detection reaction of amino groups as is possible with the ninhydrine reagent, with hydroxynaphthaldehyde, and with picric acid. In some synthetic investigations, for example, quantitative peptide couplings were stated ($> 99.5\%$) based upon negative ninhydrine or picrate tests of remaining amino functions, whereas the real final yield of crude end product considerably deviated from the calculated amount deduced from the analyitical data [137, 160].

In our group, we try to control the peptide synthesis by the aid of the centrifugal reactor (see p. 75) and flow photometry of circulating reaction liquids based upon the spectroscopic properties of the Ddz-protecting function, in combination with continuous recording of reaction curves. In the actual coupling step, however, the consumption of Ddz-amino acids by incorporation into the growing peptide on polymer does not represent a direct measure of peptide bond formation itself. Side reactions on polymer which might decrease the concentration of Ddz-derivatives in the circulating reaction liquid would affect the shape of the consumption curve in the same sense as peptide formation. Therefore, only the total incorporation of Ddz-amino acids onto polymer is in this way monitored photometrically. Integration of the reaction curves renders it possible to quantify the measurement. Each peptide coupling is repeated until the photometric level on the recording at the end of the reaction period twice shows identical values. This indicates qualitatively that no further incorporation of amino acid residues takes place on polymer. In other words, our synthesis control system indicates completed peptide reactions, but it does not provide the direct analysis for quantitative transformation of amino groups on polymer. Unfortunately, to date we have no additional test which could be integrated in the described control system to determine precisely and nondestructively these eventually remaining small amounts of functional sites. The very interesting picrate method [84] endangers the acid-labile Ddz-protecting group. It seizes other basic centers on polymer and forms an increasing optical background value which disturbs the continuously operating photometric control system [137]. More pronouncedly, the perchloric acid titration technique [134] is incompatible for the same first two arguments just mentioned. In addition, the long-lasting titrations under acidic condition saturate teflon parts of the centrifugal reactor with acetic acid. This can cause acetylation of amino groups on polymer in repeated peptide synthesis steps, since the acid slowly desorbs from construction materials of the reactor into the circulating reaction liquid. Further contradictions were discussed on p. 43. The hydroxynaphthaldehyd reagent is not reactive enough to be used as an additionally programmed step of the progressing synthesis, since the Schiff base formation is reported to need about eight hours for completion, approaching an equilibrium level dependent on the aldehyde concentration [92].

Apart from the question of compatibility of chemical tests with the continuous photometric control of a gel phase synthesis, it has to be accentuated that the monitoring of peptide couplings by any of the methods mentioned is undoubtedly essential to judge the quality of any synthetic run. Not in the least, high excesses of reagents introduced into the synthesis as well as preprogrammed reaction times do automatically guarantee coupling rates better than 90%. The reality of daily synthetic work on polymer indicates that quantitative (100%) couplings are the exception! Deviations from quantitative transformations, however, signify the formation of exponentially increasing amounts of false sequences, which, if out of control, place doubts upon the scientific value of a whole synthesis.

To determine the product distribution on polymer at the end of a Merrifield synthesis, we, like others [161–163, 212], found the Edman degradation particularly helpful. Since we developed a direct method to convert the cleaved thiazolinones into thiohydantiones without any losses caused by additional manipulations [164], hereby the sequence determinations can be quantitatively evaluated. The identification of phenylthiohydantoines with simultaneous integration is performed on a high pressure liquid chromatograph [165]. In this way the ratio between the target and the defective sequences can be quantified as well as the origin of deletions during synthesis sensitively can be detected. This analytical measure clearly shows the result of all synthetic operations on polymer before this judgement becomes uncertain because of additional formation of destructed sequences deriving from the reactions to detach the peptide from its support.

3.4 Suppression of Deletion Sequences

The course of a Merrifield synthesis on an insoluble gel phase is well defined as long as all transformations on polymer proceed with 100% yield. As described in Chapter 2 (p. 9) and as indicated in the preceding section, deviations from this goal cause the undesired formation of defectively shortened peptides, which are to be divided into "truncated" and "failure sequences" (for definition of these terms see p. 14). The statistical number of the former is much smaller, since most of them might be reincorporated occasionally into the proceeding chain elongation reactions forming the bulk of peptidic by-products, namely the "failure sequences". The starting points of these impurities accordingly are amino functions, which became aminoacylated at a wrong time of the proceeding gel phase synthesis regardless of whether these amino groups originated from deprotections or deprotonations completed too late or from insufficient peptide coupling reactions. These three types of deviations intermediately form the pool of truncated sequences, which can be defectively reincorporated into the peptide synthesis. Therefore, as long as the completion of each reaction on polymer cannot be guaranteed analytically, the only way to control the main origin of by-product formation is to exclude truncated sequences from further peptide couplings. This can be done by blocking with an acylating agent which has to be more reactive than the activated amino acid derivatives added in the peptide couplings. Some research groups, for this purpose, employ acetylation with acetanhydride [51] and acetic acid/dicyclohexylcarbodiimide or acetylimidazolide [166] with moderate efficiency,

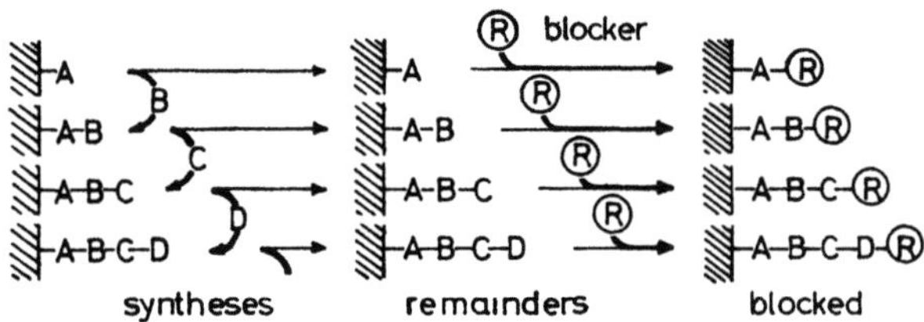

Fig. 50. The suppression of the formation of failure sequences by blocking of remaining truncated ones

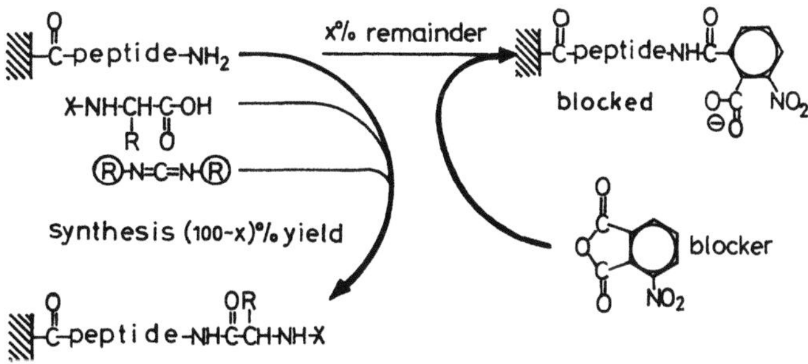

Fig. 51. The use of 3-nitrophthalic anhydride as a blocking agent in the suppression of failure sequences

however. We introduced the utilization of intramolecular anhydrides like 3-nitrophthalic- and 2-sulfonic benzoic anhydride as blocking agents [167]. These highly potent acylating compounds are rapidly aminolyzed by remaining amino functions. In this way, the majority of eventually truncated sequences becomes blocked so that the formation of failure sequences is drastically suppressed (Fig. 50).

During reaction of these blocking agents an acidic function (3-nitro-benzoyl-2-carboxylate or benzoyl-2-sulfonate) from the intramolecular anhydride bond is liberated in its salt form. All blocked defective sequences therefore additionally are marked by this acidic moiety. They can be separated from the end product by ion exchange chromatography, since the acidity of the marker function is much more pronounced than any peptide carboxylate on terminal or side [168] positions (Fig. 51). Though acetylation likewise suppresses the formation of failure sequences, it effects no facilitated separability of synthetic peptide by-products.

Today we routinely use the 3-nitrophthalic anhydride blocker: It is applicated in a single dose of a ten- to twenty fold excess on estimated amounts of remaining amino functions and added as a 0.1 molar solution in pure pyridine with ten minutes reaction time at the end of each peptide synthesis stage, in which all chemical operations are monitored by photometric control and forced to approach completion. We have no indications to date that the acidically marked, blocked sequences on polymer cause undesired side reactions in subsequent stages of the synthesis, particularly in further peptide couplings, probably because of the acidity of the blocker functions (pK ~ 2), which in their anionic salt form possess only a very diminished nucleophilicity.

Under the acidic conditions of the Edman degradation (trifluoroacetic acid/40 °C/ 45 minutes), the blocker was found to be completely cleavable from peptides on polymer containing sequences terminated by 3-nitro-2-carboxy-benzoyl. Therefore, we are in the position to determine directly on polymer the contamination of the target sequence with defective ones, though the latter originally were blocked at their N-terminus. This is done by quantitatively exploitable Edman degradation in gel phase, a technique which was already mentioned in Sect. 3.3.6.

3.5 Cleavage of Peptides from Polymer Support

All chemical operations on gel phase have to be followed by intense rinses to separate
excess reagents and soluble by-products from the polymer-supported peptide. The solvents
used for this purpose should easily dissolve the compounds to be washed-out and should
completely swell the gel phase to reach each inner section of the macromolecular coil. In
some cases the wash-out efficiency is increased by intermediate shrinking of the swollen
polymer in tert. butanol, sec. propanol, or methanol. Hereby the inner volume of the gel
that contains the swelling solvent is squeezed out by the contracting polymer structure,
an effect which is comparable to the compression of a wet sponge. Often repeated swell-
shrink operations, however, endanger the mechanical stability of the beaded gel phase,
which tends to crumble to fragments the lower its cross-linking.

The completion of the wash-out operations should be checked as carefully as the
chemical transformations on polymer, since remaining by-products would cause side reactions in subsequent steps of the synthesis. In our group, this is simply and sensitively performed by the continuously monitoring control system, which operates during the whole
Merrifield synthesis. On the recordings, completed wash-out operations after each chemical reaction on polymer are indicated by a straight zero curve, since reference and sample
flow cell of the photometric unit in this situation contain nothing but pure solvent.

This state has to be attained particularly at the end of the synthesis, before the detachment of the peptide should be performed. A sample of the polymer-supported end product
should be analyzed by quantitatively exploitable Edman degradation to determine the
quality of the synthesis before cleavage as described on p. 60.

The method of liberation of the end product from its support depends, first, on the
type of C-terminal anchor bond to the polymer and, second, on the question of whether
the synthetic material has to be released fully protected, partially masked but with free
carboxylic terminus, or entirely deprotected. These aspects are part of the strategic concept — as mentioned in the beginning — which have to be carefully considered before the
Merrifield synthesis is initiated. The conditions of deprotection of the mobile temporary
N-terminal protecting group and those to cleave masking groups from side functions of the
desired peptide are interrelated with the detachment conditions projected to liberate the
end product from its support.

3.5.1 Acidic Cleavage Conditions

3.5.1.1 Trifluoroacetic Acid/Hydrogen Bromide

In most of the published Merrifield syntheses the peptides are bound to polystyrene by
benzyl ester type of linkages, which in analogy to classical methods can be cleaved by
HBr/acetic acid. Already in the incipience period of the methodology, however, trifluoroacetic acid saturated with hydrogen bromide was found more efficient, first, because of
the increased acidity of the mixture which in addition does not cause proton catalyzed
acetylation and, second, since trifluoroacetic acid is more lipophilic than the acetic one,
it solvates much better both the polymer and the peptide to be released.

By this cleavage reagent, peptides usually are liberated fully deprotected. There are only a few exceptions in protecting groups, which resist the action of trifluoroacetic acid/ HBr, namely the nitro- and tosyl-masking on the guanido side function of arginine, N-terminal tosyl-, pyridine-4-methyloxycarbonyl (iNoc, isonicotinyloxycarbonyl [215]) and trifluoroacetyl as well as 4-nitrobenzyl ester bonds. Also benzyl thioether and S-acetamidomethyl groupings are not cleaved. Furthermore the imidazole masking functions on histidine like N^{im}-benzyl and N^{im}-2,4-dinitrophenyl as well as the N^{ind}-formyl residue on tryptophane are resistant to trifluoroacetic acid/HBr. However, since N-terminal protecting groups were developed as labile to acid as Bpoc-(biphenylisopropyloxycarbonyl) and Ddz-(α,α-dimethyl-3,5-dimethoxybenzyloxycarbonyl) there is no problem in selecting from the large scale of masking functions a set of suitable stable residues to sufficiently shield peptide side groups during synthesis, which all become liberated during detachment with trifluoroacetic acid/HBr. The releasing reaction proceeds very conveniently. The polymer-supported peptide has to be suspended in trifluoroacetic acid using a separation funnel with a fritted bottom through which a slow stream of pure HBr gas is insufflated for saturation and mixing. This operation is repeated two to three times for a maximum of 60 minutes at room temperature, intermitted by washes with trifluoroacetic acid. The individual filtrates in the meantime are evaporated. A cation scavenger like anisol is often added in 5—10% of the total volume to the first cleavage reaction mixture. The concentrated cleavage products are immediately dissolved in the smallest volume of isopropanol. With the aid of a syringe or a pipette, the solute is injected into a large volume of dry ether contained in a separation funnel whence the precipitated peptide raw product can be easily collected on a filter. Additional rinses with dry ether wash off remaining traces of acid and anisole.

Several anchor groupings were developed but scarcely used to facilitate the acidic detachment of the peptide end product. In this way, however, the differences in acid stability between the temporary N-terminal protecting group and the permament C-terminal link to the polymer support are diminished, which might effect losses of the growing peptide during synthesis. Nevertheless, one can select a set of protecting groups — as mentioned above — to shield peptide side functions completely against trifluoroacetic acid. This renders possible to liberate protected peptides with a free carboxyl terminus from modified polymers which are labilized against, for example, 50% trifluoroacetic acid in dichloromethane [169]: Tert. alkyl ester bonds to the polymer and tert. alkyloxycarbonylhydrazide [170] links were developed for this purpose.

3.5.1.2 Liquid Hydrogen Fluoride

The observation that liquid HF used for detachment of the peptide from polymer at 0 °C simultaneously cleaves all the other protecting groups from peptide side functions was considered a great advantage in the Merrifield scheme [110]. However, enthusiasm has vanished since liquid HF at 0 °C was found to cause a series of undesirable side reactions as well as even fissions on the synthetic end product: In peptides containing serine and threonine, we like others observed cleavage of the peptide backbone, α-peptide bonds between aspartic acid and serine or glycine can be converted into β-amides, and glutamic acid residues are transformed into pyrrolidones. The desulfurization of cysteine-containing peptides is a particularly disagreeable side effect of liquid HF though the mechanism of this reaction has not been investigated in detail. The extraordinary basicity of fluoride ions in non aqueous systems might effect some of the side reactions mentioned and therefore should be taken into consideration [213]. Although the HF detachment procedure is still used in several cases with good success, an increasing number of researchers mention poor yields of liberated peptides. In discussions it was stated that often about 50% of the synthetic material still sticks to the polymer support after the cleavage reaction. This effect might depend on poor solvation of the polystyrene matrix by liquid HF, so that peptides can be liberated and dissolved only from a shell zone of the beaded support.

Insufficient solvation of the macromolecular coil should be positively affected by introduction of polar functions. This is the case with benzhydrylamine polystyrene, which was particularly designed to facilitate the synthesis of peptide amides [171, 214]. Indeed, there are many procedures published in which the final HF detachment yielded directly from this specialized support peptide amides with good results. Nevertheless, it should be stated that the operations with liquid HF are most dangerous though a very helpful apparatus is available on the market to simplify the detachment procedure.

Since it became obvious that liquid HF is not at all the detachment reagent of choice (see in contrast [35, p. 380]), the search for a similarly efficient reagent resulted recently in the utilization of boron tris (trifluoroacetate) [172], which in trifluoroacetic acid is reported to smoothly liberate peptides bound to the polymer support by benzyl ester type linkages. All above-mentioned protecting groups stable against trifluoroacetic acid are simultaneously cleavable. Trifluoromethanesulfonic acid [173] and pyridinium polyhydrogenfluoride are proposed for the same purpose [174] (Fig. 52).

In summarizing, it must be realized that most of all acidic conditions to remove a synthetic peptide from its gel phase support include the possibility for undesired attacks on either protected or free peptide side functions as well as on the backbone, causing fissions and conversions also during work-up manipulations of already detached raw products. This is the case because most of the usually employed protecting principles — urethanes, esters, and ethers as well as some functional sites of a peptide such as alcoholic, thiolic, and amide side chain groups — can be involved in proton catalyzed eliminations, transesterification, transamidations, and cyclol formations, though some of these side reactions usually are rather feared under basic conditions.

Therefore, a search for chemically different principles to cleave the bond anchoring a peptide to its support and to deprotect side chain functions is recommended. In the author's opinion the final entire deprotection of a synthesized peptide should be separated from the chemical operation to release a peptide from gel phase into real solution. Either all protecting moieties should be cleaved and washed-out of the polymer prior to the detachment of the peptide or the completely shielded end product should be released

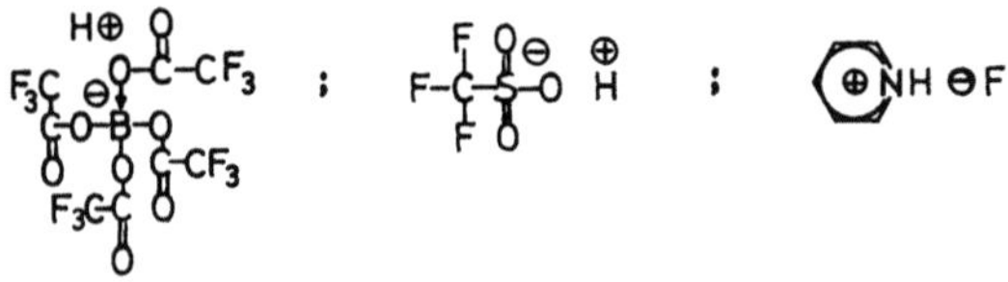

Fig. 52. Some acidic agents in the peptide detachment from polymer phase as substitutes for liquid hydrogen fluoride

from its support, so that deprotections could be performed under well-defined conditions in real solution without any interplay with the macromolecular tracery of the support. Some proposals concerning this aspect are dealt with in the following paragraphs.

3.5.2 Basic Cleavage Conditions

In analogy to the advantageous compatibility in conventional peptide chemistry of C-terminal benzyl esters with acid-labile N-protecting residues at least as sensitive as the tert. butyloxycarbonyl group, Merrifield anchored the amino acids onto polymer by benzyl ester type linkages (see p. 3).

As described in the preceding paragraphs, this bond at the end of a peptide synthesis can be cleaved under drastic acidic conditions. The more the alcoholic part of an ester withdraws electrons from this grouping, the more the ester bond is strengthened against acidic attack. For example, in the incipience period when Merrifield initially developed his method using the benzyloxycarbonyl group for N-terminal shielding, he had to fasten the C-terminal benzyl ester link to polystyrene against the deprotecting reagent HBr/acetic acid by additional nitration of the aromatic polymer. The more, however, the C-terminal anchor bond between peptide and its support is stabilized against proton catalyzed fission [175], the more this bond becomes electrophilic and should be cleavable by nucleophiles like alcohols, hydrazine, ammonia, and hydroxyl ions, yielding fully protected peptides with selectively masked or free carboxyl terminus.

3.5.2.1 Base Catalyzed Transesterification

During utilization of the Merrifield method several groups observed the transesterification tendency (see p. 41) of peptides bound to the support by the benzyl ester type of linkage in the presence of alcohols, catalyzed by tert. bases [87] (Fig. 53). The utilization of this effect to remove entirely protected synthetic end products from the support is connected with the name of Beyerman [176], who demonstrated this tender method of peptide detachment in literature. Naturally, the efficiency of the transesterification is highly dependent on the C-terminal amino acid residue linking the peptide to the support, and was found not to liberate the total amount of peptide from polymer. The average yields of the detachment are in the range of about 60—70%. The transesterification is usually performed with a 1 molar solution of triethylamine in dry methanol which shrinks the polystyrene matrix. The diminished rates of peptide removal might depend to a certain

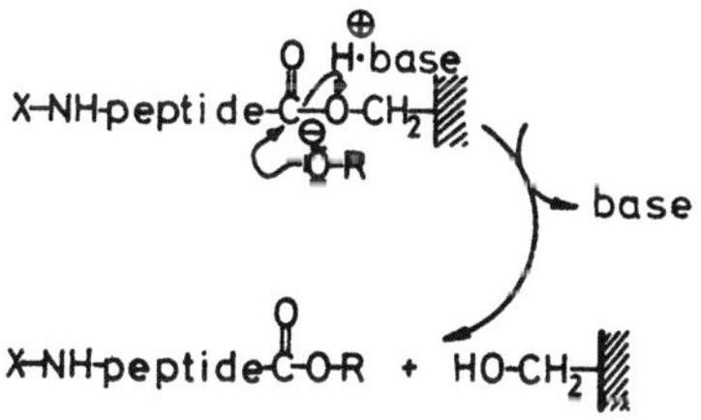

Fig. 53. The peptide release from polymer support by base catalyzed transesterification

extent on this disadvantageous state of the gel phase, though the peptide itself usually is soluble in the cleaving reagent. Solvating additives like dichloromethane or dimethylformamide are unsuitable, because of their incompatibility with the basic transesterification mixture. In combination with highly swelling low cross-linked polystyrenes we found that dilution of the cleavage reagent with one part of dioxane very efficiently increases the rate and yield of the detachment.

The presence of strong bases like triethylamine and methoxide, however, might cause up to 5% racemization at the C-terminus by direct proton abstraction. Therefore the detachment operation and the subsequent neutralization of excess reagent in the filtrates should be accomplished as quickly as possible. In our group, we particularly prefer the transesterification as a final step of the programmed synthesis in combination with the use of our continuous control system and the centrifugal reactor. The cleavage can be executed without any operational interruption of the mechanical synthesizer and can be monitored as a typical release event on the recording. The detachment operation is repeated with fresh reagent till the photometric control system attains zero which signals the end of the reaction.

The cleavage of peptides from the support by transesterification of the benzyl ester type linkage to the polymer yields the end product entirely protected, if side functions are shielded by residues stable against the basic detachment conditions. This is best fulfilled by protecting groups derived from sterically hindered tert. butanol.

C-terminal links, however, which possess an enhanced electrophilicity would facilitate the base-catalyzed method of cleavage and would even allow the use of benzyl esters to protect glutamate and aspartate carboxyl side functions. This is the case when peptides are anchored to the polymer via phenacylester bonds [68, 69] (Fig. 54). We therefore reinvestigated this type of functionalization on 0.5% cross-linked polystyrene. As already described on p. 31 this grouping can be easily introduced and reacted with amino acid caesium salts [84]. With precautions during the deprotonation steps of the first two cycles (for further details see p. 25), the peptide synthesis with dicyclohexylcarbodiimide activation on phenacyl ester-modified gel phase proceeds very smoothly. Because of its electron-withdrawing character this anchoring is further stabilized against acidic attack compared to the usual benzyl ester type of binding function. This fact, on the other side, indicates the advantageous labilization of phenacyl ester bonds against nucleophiles.

The so-called Beyerman cleavage reagent — 1 molar triethylamine in methanol — was elegantly modified in combining the two components tert. base and alcoholate into one: The reagent proposed was 2-(dimethylamino) ethanol, which is applicated as a 50% solution in dimethylformamide for self-catalyzed transesterification on polymer support [177]. Unfortunately we found this new reagent less efficient during detachment of several peptides out of benzylester type links to the polymer than with the original method. All the

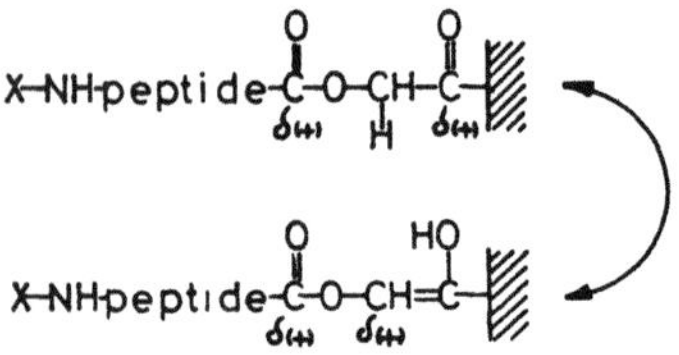

Fig. 54. The phenacetyl ester link of peptides in polystyrene is labilized against nucleophiles

more, dimethylformamide is rapidly decomposed and forms by-products like dimethyl-amides of the released peptide. It is an open question whether dioxane is a better suited solvent for this reagent. There are no results concerning the efficiency in cleaving peptide phenacyl ester bonds. Since peptides are liberated as dimethylaminoethyl esters, these products very simply can be hydrolyzed in a subsequent operation apart from the polymeric phase by addition of hydroxylic ions.

3.5.2.2 Hydrolysis

The direct hydrolysis of the benzyl ester type of linkages between peptide and polymer, because of the more drastic alkaline conditions, is a scarcely utilized method to obtain entirely protected peptides with free carboxyl terminus. We therefore modified this technique into a mild cleavage of the phenacyl ester bonds on polymer. The detachment reagent is composed of 0.5 molar triethylamine in methanol/dioxane (1 : 1, v; v) containing 5% of 0.1 N aqueous sodium hydroxide. By the programmed synthesizer, this mixture in a final step is added at 8—12 °C into the centrifugal reactor, which contains the peptide-on-polymer. Each 30 minutes the cleaving reagent is automatically renewed. The filtrates are collected in a separate flask, which — by an elegant idea of P. Fleckenstein in our group — contains solid dry ice for simultaneous neutralization of the excess bases and deep cooling of the product for final lyophilization. As indicated by our continuous photometric control system, most of the peptide is released from polymer within the first 10 minutes, whereas the support is completely discharged within 2—4 hours. To date we have no indications for C-terminal racemization caused by this method of peptide removal.

Several other types of active (enhanced electrophilic) esters to the polymer support formerly were utilized in the Merrifield synthesis to liberate the synthesized peptides by alkaline hydrolysis. However, none of these procedures to date became generally applied in practice, because of feared side reactions, catalyzed by hydroxylic ions: Deamidation, peptide bond rearrangement, cleavage of protective ester groups on side chains, attack of unmasked alcoholic functions on the peptide backbone and racemization. All these undesirable effects are not detected to be as serious in combination with the above-mentioned method of detachment from phenacyl polystyrene. In this connection it has to be accentuated that we are using side chain protections deriving from tert. butanol exclusively in combination with the acid-labile temporary N-terminal Ddz protecting group. Instead of asparagine, however, we introduce β-benzyl aspartate, which subsequent to the synthesis and the mild peptide detachment can be transformed into the β-amide by ammonolysis.

3.5.2.3 Hydrazinolysis

Suprizingly enough, hydrazine and its hydrate in diluted solutions of alcohols, dimethyl-formamide, and mixtures of both have not often been used to cleave the benzyl ester type linkage of peptides bound to the common Merrifield support, though this cleavage was found more effective than the transesterification method of detachment [178, 179]. This effect can be explained by a more sufficient solvation of the peptide on polymer by the cleaving reagent and by the distinct nucleophilicity of hydrazine. In this way fully protected peptides can be released as hydrazides, which are very suitable precursors for fragment condensations via azide couplings. The acidic conditions of the azide formation might possibly have been considered unsafely compatible with acid labile side chain protections of the peptide, e.g., masking groups like trityl or those deriving from tert. butanol. This uncertainty might partly explain the rare application of hydrazinolytic removal of peptides from polymer support. We therefore like to recommend this method of detachment, all the more since several examples are known from carefully studied conventional syntheses dealing with azide activation of peptide fragments protected by acid sensitive moieties like tert. butyl, tert. butyloxycarbonyl, biphenylisopropyloxycarbonyl, and N-S-trityl [180, 181].

During hydrazinolytic detachment of peptides, base-labile protecting groups — e.g., formyl and trifluoroacetyl — and side chain functions reactive to hydrazine, like ω-benzyl esters as well as phthalimide groupings, would be cleaved simultaneously. Although dimethylformamide commonly is used as solvent with very good success, it cannot be considered inert to hydrazine hydrate since it becomes slowly decomposed. Here, dimethylacetamide can be employed as a more stable substitute. The application of dioxane has not been tested to date, though — with reference to its very appropriate use in transesterifications and saponifications mentioned in the preceding paragraphs — it might serve as the solvent of choice also for the hydrazinolysis of benzyl ester bonds to the Merrifield polymer.

Since phenacylesters show an enhanced electrophilicity, this type of anchoring function on a polystyrene gel phase is very rapidly and quantitatively cleaved by hydrazinolysis within 2 hours at 8–12 °C. On 0.5% cross-linked supports we are applying a 10% (v) solution of hydrazine hydrate in dioxane/methanol 9:1 (v; v). The detachment filtrates are immediately neutralized with solid carbon dioxide as described in the transesterification procedure.

3.5.2.4 Ammonolysis

Ammonia is the smallest noncharged nucleophile for peptide liberation from polymer phase. Cleavage of the benzyl ester link to polystyrene by ammonolysis with the formation of peptide amides, however, liberates on the average only 60–80% of the synthetic product from the support. The best conditions for this detachment yielding directly the important class of peptide amides can be seen in the utilization of dimethylformamide/methanol 4:1 (v; v) saturated with ammonia at 0 °C, in which the peptide-on-polymer is suspended and stored for at least 3 days in a pressure bottle at room temperature.

The alcohol content of the ammonolytic reagent indicates the consequence of several observations that the amide formation seems not to proceed by direct attack of ammonia on the polymer phase ester bond but on the corresponding alkyl esters intermediately released by base-catalyzed transesterification [87]. It has to be realized that this procedure often yields a mixture of peptide amide and methyl esters, the latter of which can be transformed into amides by extended exposure to ammonia. Direct ammonolytic cleavage with liquid ammonia itself, first, is hindered by the poor solvation of the lipophilic matrix in the polar medium. Second, if remaining anchor functions are still present like chloromethyl- or bromoacetophenyl-sites, on 0.5% divinylbenzene containing supports, liquid ammonia was found to cause additional cross-linkage, reacting twice with those functional groups forming secondary amine bridges (see p. 29–30).

To date there have been hardly any experiences concerning the efficiency of ammonolytic detachment of peptides linked to the support by phenacyl groupings. For analytical purposes, in our laboratory we detected, by the aid of our photometric synthesis control system, the release of lipophilic peptide amides within 24 hours of ammonolysis. The reagent we are utilizing is 10% (w) ammonia in dioxane/methanol 9:1 (v; v), with renewal of the mixture delivered to the centrifugal reactor every 2 hours. It should be recalled, however, that in most of the cases when peptide amides are to be built up, the target sequences are synthesized on a benzhydrylamine support [171], which by HF-cleavage directly releases the entirely deprotected amide (see p. 64).

3.5.3 Modified Cleavages

Modified links between peptide and polymer support are introduced for selective fission of this anchoring bond at the end of the synthesis by detachment procedures which do not interfere with other functional sites or protective groups of the peptide.

Though several specialized anchor groups on polymer were designed and synthesized, to date none of those proposals have become generally applied in practice – as we often can observe the fact that there are advantageous reagents or reaction principles, which remain "home methods" of the inventor's laboratory.

Nevertheless, some of those modified cleavage methods will be discussed in the following: The fission of phenacyl ester links to the polymer – from several aspects already described on the preceding pages parallel to the cleavage of benzyl ester bond – can be considered a modified method of detachment depending on the enhanced electrophilicity of this grouping. In the literature particularly the thiolytic fission of this linkage is mentioned as a selective method to liberate peptide acids from polymer (Fig. 55). Yet utilizing the recommended cleaving reagent sodium thiophenolate in dimethylformamide [68, 69, 182], we experienced several side reactions of the nucleophile with the peptide Ddz-Ser(tBu)-Asp(OtBu)-Gly-Thr(tBu)-Phe-Thr(tBu)-Ser(tBu)-Glu(OtBu)-Leu-Ser(tBu)-OH (Secretine 2–11) not detectable during transesterification and alkaline hydrolysis of the same material in the detachment from phenacylpolystyrene.

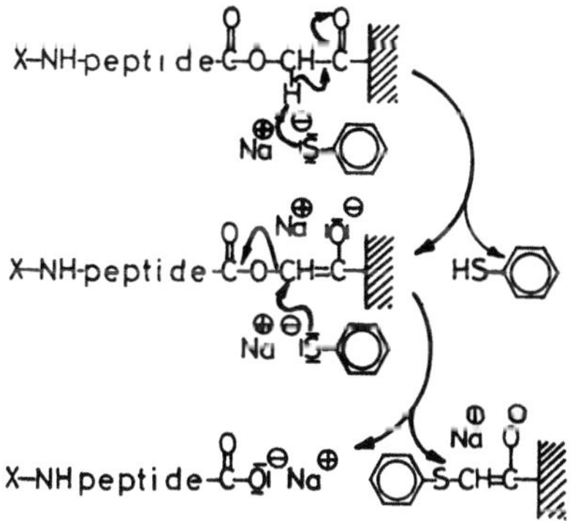

Fig. 55. The thiolytic fission of polymer phenacetyl peptide esters

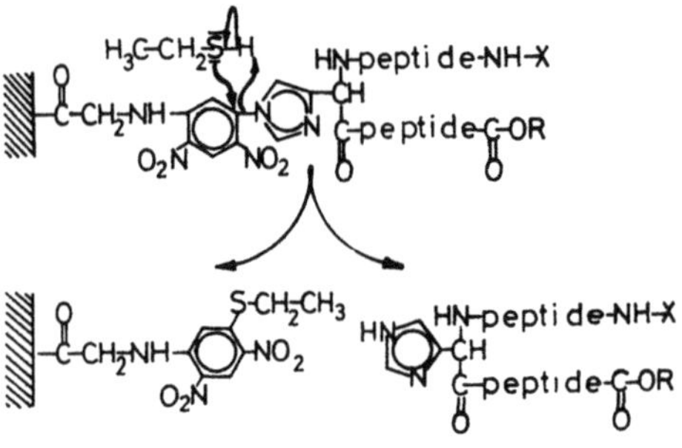

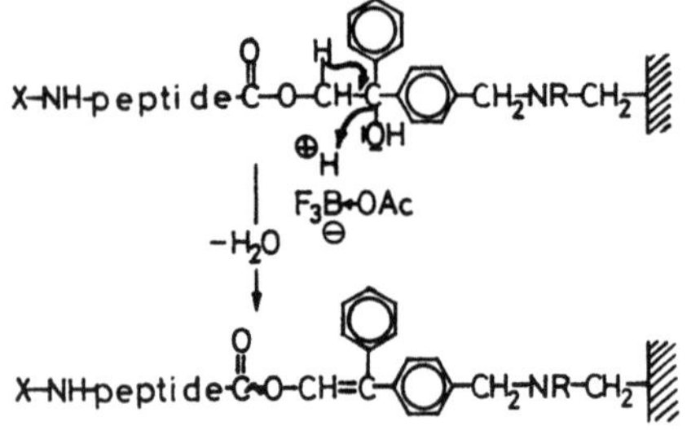

Fig. 56. The thiolytic release of peptides bound to the polymer phase via dinitrophenylene bridges on histidine

An interesting variant of peptide fixation was briefly mentioned on p. 7–8, namely the bridging technique to bind trifunctional amino acids as starting derivatives via their side chain groups onto the polymer phase [49, 50]. In this way a bidirectional peptide synthesis should be possible beginning at the amino or the carboxylic terminus, respectively. In this methodical proposal histidine was attached at its imidazole moiety by dinitrophenylene links to the gel support. At the end of a Merrifield synthesis the assembled peptide can be released under thiolytic conditions usually employed to deprotect N^{im}-dinitrophenyl-masked histidine (Fig. 56). This specific detachment operation does not endanger any other protected side groups, provided that disulfide bonds are not used to shield sulfhydryl groups of cysteinyl residues in the synthesized material. (For further discussions see [183].)

From several handle molecules proposed for subsequent activation of the anchoring link to the polymer, we mentioned a 4'-aminomethyl-2,2-diphenyl-ethane-1,2-diol-modified support for facilitated peptide release at the end of a Merrifield synthesis [142, 184] (see p. 28–29). The glycolic handle contains a primary and a tertiary alcohol function of which the former is esterified by the built-up peptide. The latter one can be eliminated through the action of trifluoroacetic acid or borontrifluoride etherate in acetic acid resulting in the formation of an activated diphenylvinyl ester on polymer (Fig. 57). This modified ester can be cleaved smoothly by all of the nucleophilic agents mentioned in the preceding paragraphs, resulting in transesterification, alkaline hydrolysis, hydrazinolysis, and ammonolysis of the subsequently labilized linkage to the gel support. To yield fully protected peptides in this way, naturally all protecting groups of side functions have to withstand the acidic reagents applied once at the end of the synthesis to activate the C-terminal ester bond. The opposite is the case for the temporary N-terminal protector, whose cleavage

Fig. 57. The subsequent elimination of water from the 4'-aminomethyl-2,2-diphenyl-ethanediol handle, which activates the peptide anchoring ester bond in the polymer phase

conditions do not have to preactivate the C-terminal handle molecule during synthesis of the peptide. The former demand is fullfilled, e.g., by protecting groups deriving from halo- and nitro-substituted benzyl alcohols, the tosyl-, acetamidomethyl- [185], trifluoroacetyl- [186] and the pyridine-4-methyloxycarbonyl residues [187, 215], the latter by the lability of the α,α-dimethyl-3,5-dimethoxybenzyloxycarbonyl-(Ddz) and the biphenylisopropyloxycarbonyl group (Bpoc).

To give an example, we utilized this principle in the synthesis of the partially protected C-terminal undecapeptide acid of the human insulin B-chain i-Noc-Gly-Glu(OH)-Arg(NO_2)-Gly-Phe-Phe-Tyr(Dcbzl)-Thr(Bzl)-Lys(Z)-Thr(Bzl)-OH. From a subsequently activated sample of polymer-supported peptide, 90% of the synthetic material was released by transesterification with N-dimethylaminoethanol in dimethylformamide within 5 hours, whereas in a parallel test from a nonactivated part, it took 70 hours for almost complete detachment of the peptide. After self-catalyzed aqueous hydrolysis of the intermediate 2-dimethylammonium ethyl ester groups and purification of the product, the peptide was obtained in 19% yield calculated over all synthetic operations including the introduction of the handle molecule onto the polymer support [188].

Among series of other handle-modified supports proposed for safe anchoring of a peptide during synthesis and facilitated detachment afterwards, the possibility of photochemical excitation of the linking moiety has to be mentioned. Peptide-3-nitrobenzyl esters bound via benzamidomethyl links to the polymer can be released as fully protected acids by irradiation for 18—24 hours at 350 nm of the gel phase suspended in methanol or ethanol. In this way the fully protected sequence of the luteinizing hormone releasing factor in its acid form was obtained after purification in 64% overall yield based on the initial load of the polymer. This method of photochemical detachment demonstrates its excellent potential, particularly in relation to the photosensitive side groups of histidine, tryptophane and tyrosine occurring in that sequence [189].

In spite of repeated investigations [190] we experienced less satisfying results in utilizing enzymatic hydrolysis to liberate synthetic peptides from polymer (see on the contrary [183, 191]).

Terminal sequences of collagen, which determine its antigeneity, were synthesized on a gelatinous polystyrene support, which was modified by hexa-Lys(Boc)-handles. Subsequent to the synthesis, the ϵ-tert. butoxycarbonyl groups were cleaved selectively before the polymer-supported peptide was incubated for 24 hours with trypsin at pH 7.9/37 °C. This procedure released as little as 2.5% of the polymer-bound collagen sequence. Parallel tests without enzyme resulted in about the same percentage of peptide, which indicates a nonspecific detachment, obviously dependent on high concentrations of hydroxyl ions in the microenvironment of the oligo-lysyl link, as a neighbor group effect of unshielded ϵ-amino functions on lysyl peptide bonds.

The catalytic hydrogenolysis of benzyl esters — a well known and clean procedure in conventional peptide chemistry — recently was applied successfully to peptide cleavage from polystyrene supports [216]. The hydrogenation of the polymeric benzyl ester anchor bonds was performed in the presence of palladium II acetate with hydrogen at 60 psi, 40 °C for 24 hours, liberating some pentapeptides in good yields [217].

4 Automatization of the Merrifield Peptide Synthesis

Already in the incipience period of Merrifield's method it was demonstrated that the new idea included the possibility for automatization of most of the chemical operations necessary for peptide synthesis [39]. On this aspect was based a great part of the enthusiasm which still accompanies the development of the methodology to date: Like a fata morgana all routine work in conventional peptide synthesis seemed to be transferable to a more or less automatically operating machine. The only remaining labor was to refill reagents and solvents, to maintain the instrument, and to reap the end product.

Yet, as described in the preceding chapters, the inherent categorical imperative of the method demanding completion and control of all chemical transformations on cross-linked polymer supports did not allow a direct breakthrough to automatic synthesis of real proteins. Though from the point of view of chemical problems there are still shortcomings in this respect, today several mechanical instruments are available on the market to perform manually or program-controlled gel-phase syntheses.

In the following sections it is not so much the author's intention to balance different scientific tools against each other, but rather to propose some criteria for the judgement of a reliable peptide synthesizer.

4.1 Components of an Automatic System

To make a complicated story short, we can state that there are no fully automatically operating peptide synthesizers on the market. Though in the meantime different approaches were elaborated to generate a chemical feed back signal deriving from incompleted reactions on polymer phase (see p. 44), none of them to date are incorporated into commercially available machines for automatic processing of the syntheses. All instruments render a programmed sequence of operations possible, like metering and delivery of fluids, agitation of the polymer phase with liquid ones, and draining of the reactor. The mechanical execution of those functions in most of the instruments are self-controlled electronically by sensor systems resulting in half-automates. These machines are capable of performing all types of chemical reactions on cross-linked polymers not only restricted to peptide syntheses. To accentuate it again, the chemical efficiency of the synthetic operations remains uncontrolled by the automate, since there is no information fed back into the preprogrammed run of the instruments.

4.1.1 Programming

The first apparatus designed as characterized above was constructed by Merrifield and Stuart [39]. For programming it contained, in the mode of a musical box, a big drum into which little pins were inserted. The pins activated microswitches which were connected to various functions of the machine. The duration of any function is determined either by a row of pins on the drum and therefore limited by the width of the latter, or by external time relays activated by single pins. This very reliable old-fashioned way of "music box programming" is restricted in its variability and length and is not always easy to survey. The additional use of time relays extends the flexibility of the program, yet includes the disadvantage of overloading of the instrument with electromechanical switches. This principle is not altered in general by substitution of "pins on a drum" by "hollows on a punched chart" as utilized in modernized programmers of the musical box type. The chart may be connected at its ends to form an endless program with fixed functions and durations of those.

Positively the possibilities of programming are bettered by introduction of electronic logics using integrated circuits. An electronically performed time base is multiplied or divided as given by the coded information from a punched tape resulting in unlimited duration of all time functions, if necessary. The programs are punched, read, and copied on teletype writers, as is common in computer techniques. In this way the necessity for electromechanical switches is nearly excluded [192].

Future instruments for programming and complete synthesis control will contain micro processors to execute the functions of the machine based on signals from the mechanical operations as well as from the chemical reactions of the synthesizer. A minimum of orders will have to be preprogrammed by dialing into an electronic memory system. The control unit will check the chemical transformations and draw logical conclusions to optimize the proceeding synthesis, as further discussed in Sect. 4.1.4.

4.1.2 Transport and Metering of Liquids

For chemical reactions on an insoluble polymer, dissolved reagents and washing media of maximal solvation power are needed to swell the cross-linked support. In peptide synthesis most of the reagents and solvents used are volatile, corrosive, and toxic. Therefore, these properties of the fluids also have to be taken in consideration during storage, pumping, and metering in a peptide synthesizer.

All reagent solutions and washing media in the Merrifield process are added to the polymer phase in high excess because of the demand for complete chemical transformations. Therefore, on the first view there is no need for highly accurate metering of fluids, the more it is necessary to avoid cross-contamination of all reagents and solvents during delivery to the reaction vessel of the synthesizer. This is usually realized in connecting the reservoir bottles to the top of a metering flask by open-ended tubing. The metering burette is positioned on a higher level than the reservoirs. By pressurizing the latter, liquids are pumped through the tubings up to the metering system. Depressurization stops the flow and causes the amount of liquids till in the line to flow back to the reservoir by gravity.

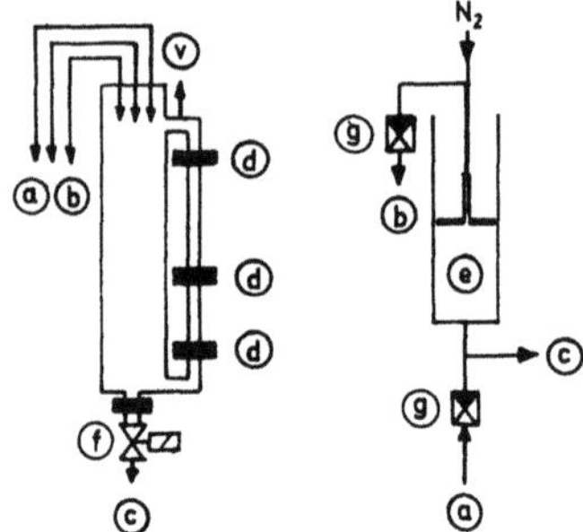

Fig. 58. The metering of liquids in different peptide synthesizers. (a) Fill line from the reservoir; (b) Reflux to the reservoir after filling; (c) Delivery line into the reactor; (d) Metering level photo sensor; (e) Variable volume by an adjustable plug; (f) Solenoid valve; (g) Check valve; (v) Vent line

The metering is performed by the aid of photo sensors variably fixed to a side arm of the metering flask. Being pumped into the latter, the reagent solutions or washing media activate the sensor in reaching its level position. The sensor signal stops the metering operation by depressurization of the reservoir. (For other types of metering liquids see [193, 194]. Though this principle of pumping and metering is not the only one conceivable, it excludes cross-contamination of chemicals and renders transport of liquids possible with a minimum contact to moving parts like valves, which have to be chemically inert to all agents in use (Fig. 58).

Yet, in view of self-controlled transformations during automatic synthesis, a future instrument will need a precise metering system of accurate reproducibility. This is because all chemical transformations then have to be monitored by additional analytical reactions following each synthetic operation. The photometric detection of the consumption of dissolved reagents and of the liberation of molecules from the polymer into the liquid phase during the proceeding synthesis also need precise metering of reagents (for further discussion see p. 77). This technical problem might be solved in a modified way of that described above [194] using individual buretts for each reagent including, however, a variably adjustable overflow principle in addition to the photosensor system, which determines exactly the desired volume of the liquid to be measured [131]. A system of syringe pumps which would combine accurate metering and transport is also conceivable.

4.1.3 The Reactor

The reaction vessel of a peptide synthesizer can be considered the heart of the instrument. Merrifield, for this purpose, introduced a shaker principle to mix the liquid and the gel phases. Others are using mechanical stirring or gas injection to agitate the reaction mixture (Fig. 59). Depending on the type of mixing (shake, stir mechanically or by gas) in the reactors, one can observe that beads of the polymer support sometimes stick to the wall of the vessel without contact to the reagent solution. Therefore, additional rinsing facilities like a rotating disc and an extra pump or lipophilic coating of the glass surface are needed to diminish this undesired effect. To separate the phases, the mentioned reactors have glass filter bottoms small in diameter relative to the volume of the vessels. Therefore precipitating microcrystals and ground polymers can easily block the filter and slow down the draining of the reactor. Though from the point of view of chemical reactivity gelatinous poly-

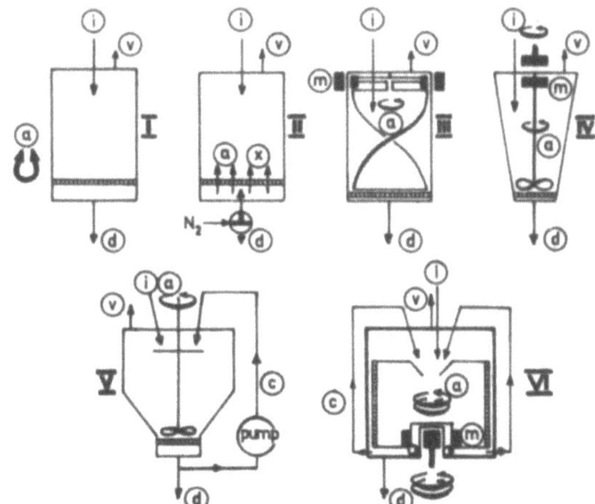

Fig. 59. Different types of reactors used in the Merrifield peptide synthesis. (a) Agitation mode; (c) External circulation; (d) Drain line; (i) Reagents inlet; (m) Magnetic clutch; (v) Vent line; (x) gas stirring

mers of the lowest degree of cross-linkage are desirable, one can hardly stir or shake those soft and mechanically labile highly swollen particles without destruction and blockage of the filter of the reactors mentioned.

For these reasons the centrifugal reactor was designed [61] to overcome the shortcomings indicated above (Fig. 60). In a thermostated glass housing a slightly smaller rotor is mounted, whose cylindrical wall consists of porous quartz. This centrifuge beaker is connected to the drive by a magnetic clutch only so that the reactor is completely closed and inert to any chemicals. Reagent solutions and solvents are delivered from the synthesizer by tubing through the screw top of the reactor. The rotor contains the polymer beads in a preswollen state. When the reactor is filled with a metered amount of liquid, rotation of the centrifuge basked is started. The centrifugal motion causes the gelatinous support inside the spinning rotor to move to the porous vertical wall of the latter and to stick there, whereas the liquid phase penetrates the polymer and the basket wall. The centrifugal force causes the liquid to climb the inner wall of the glass housing surrounding the rotor. At the inner side of the screw top of the housing, liquids are turned round to plunge back into the center of the spinning rotor, forming an intense current that rapidly and continuously penetrates the polymer inside. Since the gelatinous support sticks to the porous wall of

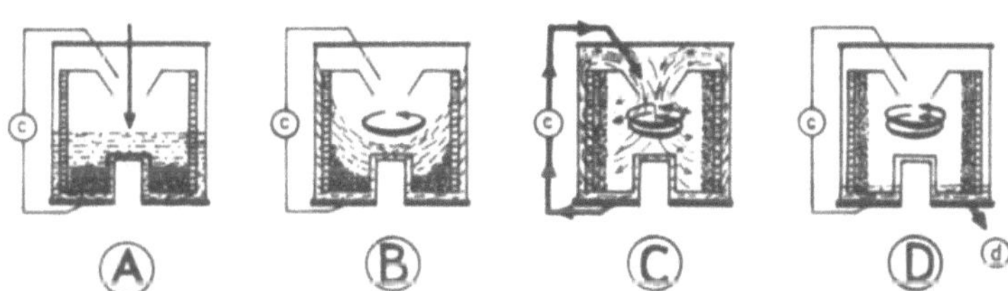

Fig. 60. The mode of operation of the centrifugal reactor. (A) Filling; (B) Start of rotation, (C) Internal and external circulation of reaction liquids caused by centrifugal forces; (D) Draining by spinning off; (c) Flow cell; d Drain line

the rotor, there is no motion of the polymer beads relative to the large cylindrical filter surface and therefore there is no abrasive action on even mechanically labile gels. None of the polymer particles can avoid contact with the liquid phase.

In addition to the intense internal current of liquid continuously penetrating the polymer support inside the spinning rotor, an external flow can be diverted since the mode of action in the vessel is comparable to a rotary pump. In this way external flow cells can be circulated with reaction solutions from the centrifugal reactor without any additional pump. Hereby, the alterations of concentrations in the circulating solutions caused by consumption of reagents or liberation of molecules cleaved from the gel support can be measured directly by flow photometry (UV, IR, conductivity, etc.) and recorded continuously. On opening an outlet valve on the bottom, the reactor is emptied rapidly and completely by spinning off.

Another advantageous effect can be seen in the fact that in the use of the spinning rotor the mass transfer of dissolved molecules in the polymer is no longer dependent on nondirected diffusion. The strongly oriented vector of the centrifugal force overcompensates by far any other power acting on mobile molecules inside and outside the macromolecular tracery of the support, which by its cross-linkage is fixed in space and sticks to the filter surface of the porous centrifuge rotor. In the use of the spinning reactor a steady state of reagent concentrations in the polymer phase can immediately be realized. Therefore, kinetics photometrically recorded from circulating reaction solutions are not falsified by a diffusion parameter. This fact is also worth mentioning in the application of the centrifugal reactor for substrate reactions with polymer-supported, immobilized enzymes.

4.1.4 Control of Functions and of the Synthesis

As already indicated in the preceding paragraphs, an automatically operating peptide synthesizer needs complet monitoring of the execution of all orders programmed, and a feedback signal from the course and efficiency of the chemical reactions on polymer.

An electronic block has to prevent the skipping of any programmed information to the machine. Any malfunction including the leak of a valve has to stop the programmed operation and activate an alarm system. After a preset duration of an alarm, draining of the reactor followed by an automatic rinse of the polymer is desirable.

The control for flow, metering, and overflow of liquids, as well as the draining of the reaction vessel is best performed by photosensors. The agitation of the reaction vessel (shake, stir, rotation) can be monitored by magnetic induction deduced from the moving part of the reactor and not from its drive.

The control of the chemical operations like deprotection, deprotonation, coupling, blocking, washing, and detachment in future instruments has to proceed continuously parallel to the operations. From the author's point of view this will only be possible with flow photometry on circulating reaction solutions, though this method lacks [95] the highest accuracy [195] desired. Since the precision [195] of photometric determinations is well established in analytical chemistry, the deviation of up to 0.4% in the measurements from the true values can be compensated by electronic comparisons of the course of kinetics in repeated reactions. In other words: The future computer control of peptide synthe-

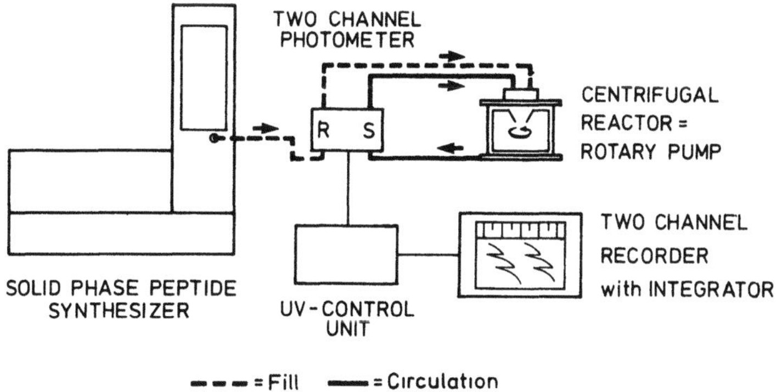

Fig. 61. The photometric synthesis control system in cooperation with the centrifugal reactor to monitor the Merrifield process

ses on gel phase needs the potential to judge the chemical operations qualitatively and quantitatively as well. Then the direct photometric monitoring of circulating reaction solutions can be performed with the necessary reliability and can be fed back to the automatic operating synthesizer.

In our laboratory this principle of photometric synthesis control is already in use, yet man's decision in judging the recording of the synthesis has to be substituted in the future by a computer system.

The utilization of the photometric synthesis control [95] was already mentioned in Sect. 3.1.3, 3.2.4, and 3.3.6 (Fig. 61). The system consists of a two-channel flow photometer with reference cell technique operating on two wavelengths. It is connected to a continuous line recorder with fifteen times automatic scale expansion at constant sensitivity and a graphic integration supplement. The photometer reference flow cell of a wide inner diameter has an optical path of 5 mm. It is inserted between the metering system of the synthesizer and the centrifugal reactor and allows a flow rate of 5 to 10 liters/hour. In contrast the sample flow cell, which is connected to the centrifugal reactor for perfusion with reaction liquids, has a narrow inner diameter and an optical path of 1 mm. This difference to the reference cell was additionally introduced to manage in part the wide range of varying concentrations to be measured in the flow-through system by optical subtraction of solvent influences with the reference technique.

All chemical operations on the recordings are indicated as kinetic curves (Fig. 62). A constant plateau means no alterations in the concentrations of the circulating reaction liquids. No further transformations take place on polymer phase. The rinses following each chemical reaction are recorded as typical wash-out events and are extended until the photometer attains zero. The control for completion of the chemical operations, in addition to the judgement of constant end values as mentioned above, is performed by repetition of the operation in question after completed washings. In the repetition, completed deprotections and, for example, finished peptide release after detachment at the end of the synthesis, are indicated by constant zero recordings. Completion of deprotonation and peptide coupling is demonstrated by curves identical to the preceding one in repeated reactions.

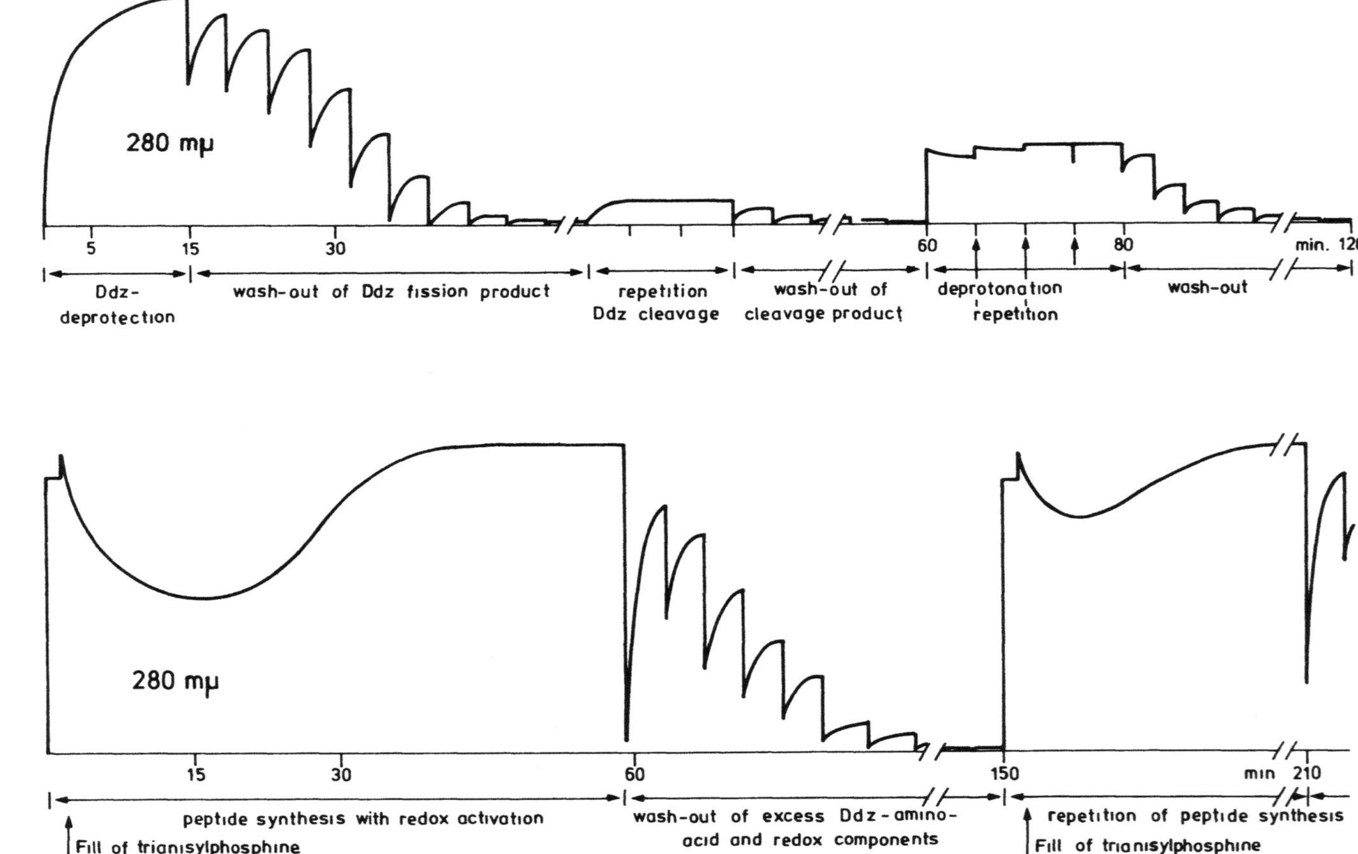

Fig. 62. The recordings from the individual operations in the Merrifield synthesis by the photometric synthesis control system depend upon the absorbance of the Ddz-amino acids used

The synthesis control by the aid of picric acid though also based on photometric measurements needs additional chemical reactions to form, first, quantitatively — and free from background reactions — the desired picrates of remaining amino functions on polymer support and, second, the complete liberation of those picrates by exchange reaction with excessively added tertiary base. Without question, there are excellent devices to solve the problem of how to technically introduce this principle of synthesis control into a fully automatically operating system. Without any doubt, the photometric determination of small amounts of picrate released from polymer can be performed with higher absolute accuracy [195] than the photometry of alterations in the total concentration of circulating reaction liquids, though in both cases the precision [195] of the determinations is identical. The contradiction in the picrate method has to be seen in the principal inaccuracy introduced by any additional chemical reaction on insoluble gel polymers. This question was discussed in Sect. 3.4.

The third method for automatic monitoring of the gel phase synthesis has long been known [196] — the perchloric acid titration on polymer — and recently was built into a computer-controlled peptide synthesizer [134]. Here, the results of the analytical determinations are fed back into the monitor system, which is programmed to optimize the course of the synthesis based on the titration data. This example of best realized automatization to date, however, is restricted to acid-tolerant synthesis strategies. The limiting chemical conditions of the analytical principle in use were already discussed on p. 43—44.

4.2 Critical Statement on Peptide Synthesizers

Automatic synthesizers are robot chemists, which must deal with toxic, volatile, and highly corrosive chemicals in a water-free system. This aspect should be taken as a general base to judge the practical value of an instrument available on the market. The flexibility of programming, the degree of self-control of functions by an alarm system, the reliability of the liquid metering system are further parameters which have to be taken in consideration to estimate the utility of a commercial peptide synthesizer. Particularly, the reaction vessels offered and the type of agitation need the critical attention of the user. It is obvious that in most of the synthesizers reactors are installed, which contain fritted glass filter surfaces, too small in diameter compared to the volume of the vessel. It is desirable to have interchangeable filter discs of different porosity to adapt the process of phase separations optimally to the type of polymer in use. If the phases are mixed by stirring or shaking, the mode of agitation of the beaded polymer in relation to the frit bottom has neither to grind the gel material nor to form dead edges in which the beads are out of permanent contact with the reaction solution. The advantageous properties in these relations of the centrifugal reactor were described on p. 75—76. They may particularly be considered on their practical merits because of the possibility of perfusing directly any external detector flow cell with reaction fluid by the mode of action of the reactor itself.

With most care, one has to look at possible corrosion on electromechanical parts like relays, multifunctional switches and on solenoid valves. In the latter relation one has to warn of a prevalent misunderstanding: Polytetrafluoroethylene, though very inert against corrosive agents, is not at all tight against diffusion of hydrogen chloride, trifluoroacetic acid, and other volatile reagents in long-lasting contact with those fluids or their vapors. Therefore, teflon parts of, for example, a solenoid valve should be mounted with an air-pad distance to the metallic drive of it. Lines connecting the reagent reservoirs and the metering system which do not need to be flexible should be made from glass tubes fixed to the housing of an instrument instead of teflon tubings. The whole cabinet of a peptide synthesizer should be manufactured from polypropylene instead of coated iron-plates. Last but not least, attention should be paid to the solutions offered to protect the whole complex system of a peptide synthesizer from atmospheric humidity.

5 Critical View on the Applicability of the Merrifield Synthesis

In this final chapter it is not the author's intention to add to the existing excellent and very comprehensive reviews of [33, 35] applications of gel phase peptide synthesis another less complete one. Instead, from a very personal view, some points are selected for discussion on the applicability of the Merrifield principle, which is of indubitable wealth, not at all only in peptide chemistry.

5.1 Aspects on Product Isolation and Purification

There is no doubt that we shall be able to closely approach the theoretical aim in the Merrifield synthesis of quantitative transformations on gel phase. However, we shall never reach this goal exactly under all varying conditions of individual multistep syntheses. Consequently each deviation which was fully discussed in Sect. 2.2 causes the accumulation of synthetic contaminations in the product.

Besides this fundamental reason one has to realize that the result of even a most accurate multistep synthesis on polymer can be questioned by the chemical reaction necessary to release the final product from its support. Parts of the target sequence can be destroyed and desired protecting groups on side functions may be cleaved. These facts furthermore multiply the possibility for contaminations. Therefore, the overall success of any Merrifield peptide synthesis is inherently related also to the quality of isolation and purification procedures (see for example [168]).

In contrast, most of the newly reported organic syntheses with polymer-supported reactants [197] are distinguished from the restrictions mentioned by increased selectivity of the reaction pathway resulting in less impure products than known from conventional transformations. This is the case, first, because of the topochemical separation of reaction locations on polymer, and second, because of the monostep nature of these syntheses. In organic preparations the products are usually isolated by crystallization. Peptides in general do not show a pronounced tendency to crystallize. Moreover, this classical technique of purification in the peptide field does not per se guarantee chemical homogeneity, since peptides tend to form mixed crystals with similar sequences.

Most of the refinement problems on peptides obtained from polymer phase are characterized by the molecular similarity of the main product and its contaminations. It is the exception that the impurities are of higher molecular weight than the target sequence. Those exceptions may result from branching, because of side reactions on unsufficiently masked reactive side groups, e.g., lysine, arginine or histidine, and from undesired di- or oligomerization by disulfide formation of cystein containing sequences. Molecular sieving of the crude products by gel chromatography on adequate Sephadex types, therefore, is usually employed to desalt the material and to pool a main fraction of low resolution as a first operation in the purification procedure. Much more effective are separations based on differences in the ionical charges of the peptides to be refined. However, ion exchange chromatography and electrophoresis are techniques which need to a certain extent water solubility of the peptide in question. As discussed in Sect. 3.3.5, a newer concept in the Merrifield synthesis deals with the intermediate preparation of entirely protected fragments on polymer of the desired target sequence. These partial sequences masked by lipophilic groupings maintain solubility in nonaqueous solvents. In those cases, the purification of the desired peptides by preparative high-performance liquid chromatography on silica gel at elevated pressure was recently demonstrated to be the most powerful preparative separation technique to date [198]. The combination of three chromatographic factors — adsorption, partition, and ion exchange — effects very high resolutions in the elution profiles.

The situation discussed in relation to smaller peptides differs from the problem of the isolation of biologically active peptides and proteins of more than fifteen amino acid residues of chain length readily synthesized on gel phase. The increased molecular similarities between the target product and possible contaminations in those cases are enlarged by the biological ones. This has to be taken into consideration in discussing the efficiency of so-called functional methods of purification [35]. These are chromatographic procedures utilizing noncovalent biochemical interactions of the synthetic molecules in question to others anchored to an appropriate polymer support. In an expanded sense of the term this technique can be named affinity chromatography. Typical examples of noncovalent relations of structures are antigen/antibody and enzyme/substrate or — /coenzyme interactions. The binding of hormones to receptor structures and aggregations like protein/lipids or -/saccharides are other examples. Since the affinity method of purification allows the separation of molecules based on biochemical functionality, contaminating peptides of, for example, a synthetic enzyme, which may have a perfect sequence in the biologically active region of the molecule but deletions outside this area, will be recognized by the affinity counterpart like the correct structure itself. One has to be aware of this effect: Though substantial amounts of synthetic impurities by this method of refinement can be safely be separated from the biologically active fraction, the latter need not be chemically homogeneous. Yet it should be stressed that this principal restriction also stands for conventionally synthesized peptides. In widening the technique, there are functional purifications in which for the chromatographic separation sequential portions of the molecules are utilized for noncovalently binding to the polymer-supported counterpart, which are not identical to the virtual biologically active site of the molecule. For example, one may generate antibodies in rabbit against a human peptide hormone for the purpose of affinity-chromatographic purification of the synthesized identical material that binds to the polymer-supported antibodies isolated from rabbit serum. In such an example the antigeneity determining portion of the sequence most probably differs from the receptor binding site of the hormone. In this way independent types of functional interactions can be utilized to increase the selectivity of the purification technique. In the future this method more and more will help to isolate the desired functional peptides in pure form.

5.1.1 Problems on Identification of Impurities

In a synthetic peptide mixture it is not always a simple task to identify a proper structure among contaminating peptides, if they are molecularly similar as is the case in preparations from polymer phase syntheses and if there are no authentic samples for analytical comparison by the aid of thin layer chromatography, electrophoresis, or spectroscopic techniques. This problem arises particularly in gel phase syntheses of fully protected peptides, which may have lost any protecting groups on side functions. Such a peptide has to be considered a contamination though it has the correct sequence. This kind of molecular deviation will be detected neither by amino acid analysis nor by sequence degradation. In addition, because of the sequence identities it is difficult to separate the impurity chromatographically. For structural identification, the resolution of those mixtures by the aid of analytical high-pressure liquid chromatography on silica gel can be recommended in combination with mass spectrometric thermalytic determinations of the protecting groups fission products. ^{1}H-NMR spectroscopy in this relation is also very helpful.

A similar problem associated with synthetic peptides of chain length > 5 amino acid residues is the identification of side reactions. Those involve substitutions on imidazole-, phenol- and indole moieties of histidine, tyrosine, and tryptophane as well as conversions, transamidation, and cyclizations of aspartic and glutamic side chains. As long as structural variations are stable under the reaction conditions of an Edman degradation, they can be detected from the proper phenylthiohydantoines in combination with ^{1}H-NMR- and mass spectrometry. Quantitative amino acid analyses of impure peptides after acidic total hydrolysis do not indicate those structural deviations between main product and contaminations.

Because of this uncertainty, we try, first, to control the course of the synthesis as completely as possible during its processing. Second, we try to determine the quality of the synthetic end product on polymer by the aid of Edman degradation with quantitative exploitation of the phenylthiohydantoines obtained [161, 164]. In this way contaminations of the product by false sequences can be detected in relative amounts of as small as 0.1% of the main chain. Generally we experienced purer peptides synthesized than liberated from polymer by any detachment reaction. This can be demonstrated qualitatively by the aid of thin layer chromatograms of the crude peptide products released from the support after the cleavage reaction and by end-group determinations before and after peptide detachment.

5.1.2 The Demand for Analytical Investigations

Because of the reasons discussed above, the homogeneity of purified synthetic peptides has to be established as much as possible by independent analytical tests to characterize the chemical identity of the product [31]. Those investigations can be distributed on two groups of end-products, namely fully masked peptide fragments A and entirely deprotected sequences or proteins B. In the former group A one has to deal with lipophilic structures not soluble in water and — because of the cleavage characteristics of protecting groups utilized — sensitive to acidic, basic, or redox agents. In the latter group B the peptides usually exhibit water solubility or at least hydrophilicity so that analytical procedures performed in aqueous systems can be employed.

Group A. Homogeneity can be demonstrated by thin layer and high pressure liquid chromatography [198, 218] on a variety of mobile and stationary phases. Supporting techniques are mass spectrometric thermolysis with detection of protecting group fission products and NMR spectrometry. The exhibition of ionic charges by selective and groupwise deprotections can by analyzed on paper electrophoresis. The compound demasked in this way can be degraded by the Edman procedure (also in combination with the dansyl end-group detection) to prove the accuracy of the sequence. Naturally an amino acid analysis qualitatively and quantitatively has to agree with the peptide structure, though it is not sensitive enough to discriminate impurities similar in composition.

Group B. Homogeneity of free peptides can be analyzed by paper and gel electrophoresis in addition to end-group analyses, thin layer and high pressure liquid chromatography [219]. Edman degradation and enzymatic digest establish the identity of the primary structure. Eventually mass spectrometric sequence determination can be used, though mass spectrometry in this group of peptides and smaller proteins is generally less helpful, because of the necessary derivatization of the material to be investigated, which may cause additional contamination. Minor impurities are not detectable as it is the case with NMR spectrometry and spectroscopic techniques like UV, IR, ORD and CD. Those methods can be employed to characterize a substance, but they do not prove the chemical homogeneity of the peptide structures. In group B peptides, bioassays based on antibody specificity, enzymatic or hormonal activities and others should be employed to demonstrate the expected biological properties of the synthetic products, though — again — these tests alone cannot indicate the homogeneity of the peptides.

5.1.3 Synthetic Peptides for Pharmaceuticals

The points discussed above characterize the problems in the syntheses of biologically active homogeneous peptides. As long as the synthetic products are prepared to detect or discover active principles of structures and to develop molecular variations potentiating known biological properties of a sequence, bioassays in vitro and in vivo on test animals are not severely limited by minor inhomogeneities of the peptides. Yet the fundamental question arises about the general possibility to applicate peptides synthesized on gel phase in man as a remedy. In this case the homogeneity of the preparation has to be considered the essential demand, because of the possible immunogenicity of each contamination and deviations in biological activities, which may even have inhibitory effects [220]. In spite of all analytical efforts to indicate the chemical identity there will remain a preparative uncertainty in peptides built up stepwise on an insoluble support.

Therefore, peptides for medical purposes should be constructed from fragments up to the size of decapeptides, of which each, after release from the polymer phase, can be purified and analytically characterized before fragment condensations are performed on gel supports to build the biologically active molecule. This concept of synthetic strategy convinces more and more of the researchers engaged in the utilization of the Merrifield principle, since only in this way can it be guaranteed that contaminations differ in their molecular weight to an extent which allows the perfect separation of impurities.

The restrictions concerning the homogeneity of synthetic peptides as well as the problem of its proof are not exclusively related to Merrifield's method of preparation. Most of the investigations in synthetic peptide chemistry to a great extent are characterized by the efforts for identification and purification of the desired structures. However, there are several unambiguous examples in which peptides are prepared, both by conventional methods and on gel phase, which would be qualified for drugs, e.g., the antamanides [199]. Here, the final preparation of the end product includes a chemical principle of selection of the proper molecules, since the target peptide is formed by cyclization of a linear decapeptide precursor. The cyclic molecule differs markedly in its chromatographic behavior from all other contaminating materials and tends to crystallize spontaneously. Similarly the matters can be described in the syntheses of oxytocins and gramicidin S.

Yet, synthetic projects in which the whole scale of trifunctional amino acids are applied usually are characterized by the need for extensive modifications of Merrifield's preparative proposals to yield the target peptide repetitively in distinct quality. This hurdle has to be overcome to apply for the legal registration of the product [200] as a medication.

5.2 Outlook on Disciplines Related to Peptide Chemistry

Nowadays, the Merrifield synthesis among all the other methods in peptide chemistry is accepted as one of the most fruitful developments in natural sciences. The idea of immobilizing one of the reaction partners influenced strongly the methodology in organic, analytical, and biochemistry in all its branches. While in the first decade of the utilization of Merrifield's proposals mainly polystyrene was taken for the support material, recently more and more knowledge from polymer chemistry was fed back into the field of polymer-supported reagents [197]. The macromolecular carriers are better adapted to the chemical properties of the immobilized residues and to the optimal reaction conditions of the individual synthetic or analytical systems: For example, special polyacrylamides are developed for peptide synthesis [65], nylon types are utilized for enzyme immobilization [201], and proteins can be anchored onto glass beads for analytical sequence determinations [202]. Polystyrene seems to remain the polymeric basis supporting typical lipophilic organic reagents. As a direct consequence of the developments in peptide syntheses on gel phases, the incremental condensation of oligonucleotides on polymer supports continues to be elaborated [203], though in this area the problems of the selective protection of side functions and of incomplete transformations are even more delicate than in peptide chemistry [204, 221].

Very generally, the researchers from different disciplines experienced the gelatinous state of low cross-linked polymeric supports not to be only physically a fourth state of aggregation, but especially a chemically useful reactive one. Future developments in gel phase chemistry will show an approximation between theoretical discussions concerning biochemical phenomena proceeding in the gel state of living matter in cells and preparative chemical utilizations. With this increasing knowledge we shall be able to realize the synthesis of definite proteins in useful quantities and of reliable quality.

6 Conclusion

In this book the author tried to balance the advantageous aspects of the Merrifield synthesis against remaining insufficient ones from his very personal point of view.

Without any doubt there are scientific questionings in peptide chemistry which can be answered by conventional methods and by the gel phase approach as well. It was demonstrated that most of the classic methodical arsenal in peptide chemistry can be successfully transferred to investigations utilizing polymeric phases, if the chemical heterogeneity of the macromolecular reaction points is taken in consideration with critical care.

As a very general judgement on Merrifield's principle it can be stated that the speed of — more or less — automatically performed syntheses on polymer supports has to be paid for by stronger efforts for subsequent purifications of the desired products and by a remaining uncertainty concerning the chemical identity of the synthetic material. Therefore, the advantages of the Merrifield method mainly are valid in those biochemical investigations where the homogeneity of the prepared peptides or proteins is not absolutely essential. Examples are the analytical studies of biological processes by structure-function relations with modified synthetic hormones and with altered active sites of enzyme sequences prepared on gel phase.

Syntheses on polymer support are of invaluable merit in the preparation of linear structures of about decapeptide size — e.g., the bradykinins — and of precursors for cyclic structures as in the case of gramicidin S, oxytocin, vasopressin, and the antamanides.

The author particularly perceives a steady challenge to organic chemists arising from still-existing shortcomings in methodical details of the Merrifield synthesis. We can feel sure that this permanent provocation will generate many excellent scientific contributions, not only to improve the results in peptide chemistry. From a retrospective view in future days we shall state that all the troubles we shouldered were worthwhile.

7 References

1. Fischer, E., Otto, E.: Ber. dtsch. Chem. Ges. *36*, 2106 (1903).
2. Bergmann, M., Zervas, L.: Ber. dtsch. Chem. Ges. *65*, 1192 (1932).
3. Wieland, Th., Bernhard, H.: Liebigs Ann. Chem. *572*, 190 (1951).
4. Wieland, Th., Schäfer, W., Bokelmann, E.: Liebigs Ann. Chem. *573*, 99 (1951).
5. Boissonnas, R. A.: Helv. Chim. Acta *34*, 874 (1951).
6. Vaughan, J. R.: J. Amer. Chem. Soc. *73*, 3547 (1951).
7. DuVigneaud, V., Ressler, C., Swan, J. M., Roberts, C. W., Katsoyannis, P. G.: J. Amer. Chem. Soc. *76*, 3115 (1954).
8. Schwyzer, R., Sieber, P.: Nature *199*, 172 (1963).
9. Meienhofer, J., Schnabel, E., Brenner, H., Brinkhoff, O., Zabel, R., Sroka, W., Klostermeyer, H., Brandenburg, D., Okuda, T., Zahn, H.: Z. Naturforsch. *18b*, 1120 (1963).
10. Merrifield, R. B.: Fed. Proc., Fed. Amer. Soc. Exp. Biol. *21*, 412 (1962).
11. Merrifield, R. B.: Endeavour *24*, 3 (1965).
12. Merrifield, R. B.: Science *150*, 178 (1965).
13. Merrifield, R. B.: In: Hypotensive Peptides (eds. E. G. Erdös, N. Back, F. Sicuteri), pp. 1–13. Berlin, New York: Springer 1966.
14. Merrifield, R. B.: Recent Progr. Horm. Res. *23*, 451 (1967).
15. Merrifield, R. B.: Sci. Amer. *218*, 56 (1968).
16. Merrifield, R. B.: Advan. Enzymol. *32*, 221 (1969).
17. Merrifield, R. B.: J. Amer. Med. Ass. *210*, 1247 (1969).
18. Merrifield, R. B.: Intra-Sci. Chem. Rep. *5*, 183 (1971).
19. Merrifield, R. B.: Harvey Lect. *67* (1973).
20. Merrifield, R. B.: In: Chemistry of Polypeptides (ed. P. G. Katsoyannis), pp. 335–361. New York: Plenum 1973.
21. Ferriere, N.: Sci. Progr., Nature *3372*, 127 (1966).
22. Halstrøm, J.: Dan. Kemi *48*, 59 (1967).
23. Suzuki, K., Ando, T.: Yuki Gosei Kagaku Kyokai Shi *92*, 696 (1967); Chem. Abstr. *68*, 50006 (1968).
24. Izdebski, J., Drabarek, S.: Wiad. Chem. *22*, 35 (1968).
25. Okuda, T.: Naturwissensch. *55*, 209 (1968).
26. Vesa, V. S.: Usp. Khim. *37*, 246 (1968).
27. Stewart, J. M., Young, J. D.: In: Solid Phase Peptide Synthesis. San Francisco, California: Freeman 1969.
28. Losse, G., Neubert, K.: Z. Chem. *10*, 48 (1970).
29. Marglin, A., Merrifield, R. B.: Annual Rev. Biochem. *39*, 841 (1970).
30. Marshall, G. R., Merrifield, R. B.: In: Biochemical Aspects of Reactions on Solid Supports (ed. G. R. Stark), pp. 111–169. New York: Academic Press 1971.
31. Wünsch, E.: Angew. Chem. *83*, 773 (1971); ibid. Int. Ed. Engl. *10*, 786 (1971).
32. Yajima, H.: Yuki Gosei Kagaku Kyokai Shi *29*, 27 (1971).
33. Meienhofer, J.: In: Hormonal Proteins and Peptides (ed. C. H. Li), Vol. 2, pp. 45–267. New York: Academic Press 1973; and also Chem. Technol., p. 242 (1973).

34. Sheppard, R. C.: In: Peptides 1971 (ed. H. Nesvadba), pp. 111–125. Amsterdam: North-Holland Publisher 1973.
35. Erickson, B. W., Merrifield, R. B.: In: The Proteins (eds. H. Neurath, R. L. Hill), III. Ed., Vol. 2, pp. 255–527. New York: Academic Press 1976.
36. Spirin, A. S., Gavrilova, L. P.: The Ribosomes. Berlin–Heidelberg–New York: Springer 1969.
37. Katchalski, E., Patchornik, A., Fridkin, M.: J. Amer. Chem. Soc. 87, 4646 (1965).
38. Wieland, Th., Birr, Chr.: Angew. Chem. 78, 303 (1966); ibid. Int. Ed. Engl. 5, 310 (1966).
39. Merrifield, R. B., Stewart, J. M.: Nature 207, 522 (1965).
40. Young, G. T.: Collect. Czech. chem. Comm. (Special Issue) 24, 33 (1959).
41. Wünsch, E.: In: Houben-Weyl, Methoden der Organischen Chemie, Synthese von Peptiden (ed. E. Müller), Vol. 1, pp. 35–39. Stuttgart: Thieme 1974.
42. Letsinger, R. L., Kornet, M. J.: J. Amer. Chem. Soc. 85, 3045 (1963).
43. Wünsch, E., Drees, F.: Chem. Ber. 99, 110 (1966).
44. König, W., Geiger, R.: Chem. Ber. 103, 788 and 2024 (1970).
45. Krumdieck, C. L., Baugh, C. M.: Biochem. 8, 1568 (1969).
46. Meienhofer, J., Jacobs, P. M., Godwin, H. A., Rosenberg, J. H.: J. Org. Chem. 35, 4137 (1970).
47. Sklyarov, L. Y., Shashkova, J. V.: J. Gen. Chem. USSR 39, 2714 (1969).
48. Meienhofer, J., Trzeciak, A.: Proc. Nat. Acad. Sci. U. S. 68, 1006 (1971).
49. Glass, J. D., Talansky, A., Grzonka, Z., Schwartz, J. L., Walter, R.: J. Amer. Chem. Soc. 96, 6476 (1974).
50. Glass, J. D., Schwartz, J. L., Walter, R.: J. Amer. Chem. Soc. 94, 6209 (1972).
51. Merrifield, R. B.: J. Amer. Chem. Soc. 85, 2149 (1963).
52. Bayer, E., Hagenmaier, H., Jung, G., König, W.: In: Peptides 1968 (ed. E. Bricas), pp. 162–170. Amsterdam: North Holland Publisher 1968.
53. Birr, Chr.: Contribution to "Euchem-Conference on Metalproteins", Helgoland 1969; unpublished.
54. Bayer, E., Eckstein, H., Hägele, K., König, W. A., Brüning, W., Hagenmaier, H., Parr, W.: J. Amer. Chem. Soc. 92, 1735 (1970).
55. Baas, J. M. A., Beyerman, H. C., van de Graaf, B., de Leer, E. W. B.: In: Peptides 1969 (ed. E. Scoffone), pp. 173–176. Amsterdam: North Holland Publisher 1971.
56. Wünsch, E., Wendlberger, G.: Chem. Ber. 105, 2508 (1972).
57. Wünsch, E., Jaeger, E., Deffner, M., Scharf, R., Lehnert, P.: Chem. Ber. 105, 2515 (1972); Wünsch, E.: Naturwissensch. 59, 239 (1972).
58. Yamashiro, D., Li, C. H.: J. Amer. Chem. Soc. 95, 1310 (1973).
59. Vollmert, B.: Polymer Chemistry, p. 322. Berlin–Heidelberg–New York: Springer 1973.
60. Vollmert, B.: ibid. p. 498.
61. Birr, Chr., Lochinger, W.: Synthesis 1971, 319; Chr. Birr, Patent No. 2017351 (1974), Munich.
62. Bayer, E., Mutter, M.: Nature 237, 512 (1972).
63. Braun, D., Cherdron, H., Kern, W.: Praktikum der Makromolekularen Organischen Chemie, p. 174. Heidelberg: Hüthig 1966.
64. Atherton, E., Sheppard, R. C.: In: Peptides 1974 (ed. Y. Wolman), pp. 123–128. New York: Wiley 1975.
65. Atherton, E., Clive, D. L. J., Sheppard, R. C.: J. Amer. Chem. Soc. 97, 6584 (1975).
66. Atherton, E., Bridgen, J., Sheppard, R. C.: FEBS Letters 64, 173 (1976).
67. Pepper, K. W., Paisley, H. M., Young, M. A.: J. Chem. Soc., 1953, 4097.
68. Weygand, F., Obermeier, R.: Z. Naturforsch. B23, 1390 (1968).
69. Mizoguchi, R., Shigezane, K., Takamura, N.: Chem. Pharm. Bull. 18, 1465 (1969).
70. Frank, H., Hagenmaier, H.: Tetrahedron 30, 2523 (1974).
71. Wehrse, M.: Doctoral Thesis, University of Heidelberg, 1976.
72. Feinberg, R. S., Merrifield, R. B.: Tetrahedron 30, 3209 (1974).
73. Bergmann, M., Zervas, L.: Ber. dtsch. Chem. Ges. 66, 1288 (1933).
74. Stelakatos, G. C., Paganou, A., Zervas, L.: J. Chem. Soc. [C], p. 1191 (1966).
75. Sheehan, J. C., Daves, G. D.: J. Org. Chem. 29, 2006 (1964).
76. Losse, G., Berndsen, G.: Liebigs Ann. Chem. 715, 204 (1968).
77. Wang, S. S., Merrifield, R. B.: J. Amer. Chem. Soc. 91, 6488 (1969).
78. Wolters, E. T. M., Tesser, G. I. and Nivard, R. J. F.: J. Org. Chem. 39, 3388 (1974).

References

79. Fleckenstein, P.: Doctoral Thesis, University of Heidelberg, 1974.
80. Wieland, Th., Birr, Chr., Fleckenstein, P.: Liebigs Ann. Chem. *756*, 14 (1972).
81. Birr, Chr., Fleckenstein, P.: will be published elsewhere.
82. Rich, D. H., Gurwara, S. K.: J. Chem. Soc., Chem. Commun. *1973*, 610; J. Amer. Chem. Soc. *97*, 1575 (1975); Tetrahedron Letters *1975*, 4037.
83. Sheppard, R. C.: In: Peptides 1971 (ed. H. Nesvadba), pp. 111–125. Amsterdam: North Holland Publishers 1973.
84. Gisin, B. F.: Helv. Chim. Acta *56*, 1476 (1973).
85. Birr, Chr.: unpublished
86. Loffet, A.: Int. J. Protein Res. *3*, 287 (1971).
87. Bodansky, M., Sheehan, J. T.: Chem. Ind. *1966*, 1597.
88. Staab, H. A., Wendel, K.: Chem. Ber. *96*, 3374 (1963).
89. Sheehan, J. C., Hess, G. P.: J. Amer. Chem. Soc. *77*, 1067 (1955).
90. Birr, Chr.: unpublished.
91. Dorman, L. C., Britton, E. C.: Tetrahedron Lett. *1969*, 2319.
92. Hirt, J., de Leer, E. W. B., Beyerman, H. C.: In: The Chemistry of Polypeptides (ed. P. G. Katsoyannis), pp. 363–387. New York: Plenum Press 1973.
93. Birr, Chr.: Liebigs Ann. Chem. *763*, 162 (1972).
94. Birr, Chr.: Liebigs Ann. Chem. *1973*, 1652–1662.
95. Birr, Chr.: In: Peptides 1974 (ed. Y. Wolman), pp. 117–122. New York: Wiley 1975.
96. McKay, F. C., Albertson, N. F.: J. Amer. Chem. Soc. *79*, 4686 (1957); Anderson, G. W., McGregor, A. C.: ibid. *79*, 6180 (1957).
97. Erickson, B. W., Merrifield, R. B.: J. Amer. Chem. Soc. *95*, 3750 (1973).
98. Erickson, B. W., Merrifield, R. B.: ibid. *95*, 3757 (1973).
99. Goodman, M., Felix, A. M., Deber, C. M., Brause, A. R., Schwartz, G.: Biopolymeres *1*, 371 (1963).
100. Wünsch, E., Z. Naturforsch. *22b*, 1269 (1967).
101. Sieber, P., Iselin, B.: Helv. Chim. Acta *51*, 622 (1968).
102. C. H. Boehringer Sohn, Abt. Chemikalien, 6507 Ingelheim/Rhein, F. R. G.
103. Birr, Chr., Ueki, M., Frank, R.: In: Peptides: Chemistry, Structure, Biology (eds. R. Walter, J. Meienhofer), pp. 409–417. Ann Arbor, U. S. A.: Ann Arbor Science Publ. 1975.
104. Birr, Chr.: In: Peptides 1974 (ed. Y. Wolman), pp. 381–384. New York: Wiley 1975.
105. Ramachandran, J., Li, C. H.: J. Org. Chem. *27*, 4006 (1962).
106. Bergmann, M., Zervas, L.: J. Biol. Chem. *111*, 245 (1935).
107. Veber, D. F., Milkowski, J. D., Denkewalter, R. G., Hirschmann, R.: Tetrahedron Lett. *1968*, 3057.
108. Wünsch, E.: In: Houben-Weyl: Methoden der Organischen Chemie, Synthese von Peptiden (ed. E. Müller), Vol. 15/1, pp. 792–796. Stuttgart: Thieme 1974.
109. Akabori, S., Sakakibara, S., Shimonishi, Y., Nobuhara, Y.: Bull. Chem. Soc. Jap. *37*, 433 (1964).
110. Sakakibara, S., Shimonishi, Y., Kishida, Y., Okada, M., Sugihara, H.: Bull. Chem. Soc. Jap. *40*, 2164 (1967).
111. König, W., Geiger, R.: Chem. Ber. *103*, 2041 (1970).
112. Birr, Chr.: unpublished
113. Kessler, W., Iselin, B.: Helv. Chim. Acta *49*, 1330 (1966); see also Faulstich, H., Trischmann, H., Wieland, Th.: Tetrahedron Lett. *1969*, 4131.
114. Goldberger, R. F., Anfinsen, C. B.: Biochemistry *1*, 401 (1962).
115. Chou, F. C. H., Chawla, R. K., Kibler, R. F., Shapira, R.: J. Amer. Chem. Soc. *93*, 267 (1971).
116. Loffet, A., Dremier, C.: Experientia *27*, 1003 (1971).
117. Reid, R. E.: J. Org. Chem. *41*, 1027 (1976).
118. Williams, M. W., Young, G. T.: J. Chem. Soc. *1964*, 3701.
119. Hünig, S., Kiessel, M.: Chem. Ber. *91*, 380 (1958).
120. Anderson, G. W., Zimmerman, J. E., Callahan, F. M.: J. Amer. Chem. Soc. *88*, 1338 (1966); ibid. *89*, 5012 (1967).
121. Blake, J., Li, C. H.: Int. J. Peptide Protein Res. *5*, 123 (1969).
122. Determann, H., Wieland, Th.: Liebigs Ann. Chem. *670*, 136 (1963).
123. Lunkenheimer, W., Zahn, H.: Liebigs Ann. Chem. *740*, 1 (1970).

124. Rothe, M., Mazánek, J.: Angew. Chem. *84*, 290 (1972); Angew. Chem. Int. Ed. Engl. *11*, 293 (1972).
125. Gisin, B. F., Merrifield, R. B.: J. Amer. Chem. Soc. *94*, 3102 (1972).
126. Brunfeldt, K., Christensen, T., Roepstorff, P.: FEBS Lett. *22*, 238 (1972).
127. Khosla, M. C., Smeby, R. R., Bumpus, F. M.: In: Chemistry and Biology of Peptides (ed. J. Meienhofer), pp. 227–230. Ann Arbor, U. S. A.: Ann Arbor Sci. Publ. 1972.
128. Wieland, Th., Birr, Chr.: In: Organic Chemistry Series 2, Amino Acids Peptides and Related Compounds (eds. H. N. Rydon, D. H. Hey), Vol. 6, p. 185. London: Butterworth 1976.
129. Kaiser, E., Colescott, R. L., Bossinger, C. D., Cook, P. J.: Anal. Biochem. *34*, 595 (1970).
130. Esko, K., Karlsson, S., Porath, J.: Acta Chem. Scand. *22* 3342 (1968).
131. Brunfeldt, K., Roepstorff, P., Thomsen, J.: Acta Chem. Scand. *23*, 2906 (1969).
132. Losse, G., Ullrich, R.: Z. Chem. *11*, 346 (1971).
133. Klostermeyer, H., Schwertner, E.: In: Peptides 1972 (eds. H. Hanson, H. D. Jakubke), pp. 108–111. Amsterdam: North Holland Publishers 1973.
134. Villemoes, P., Christensen, T., Brunfeldt, K.: Hoppe-Seyler's Z. Physiol. Chem. *357*, 713 (1976).
135. Hancock, W. S., Battersby, J. E., Harding, D. R. K.: Anal. Biochem. *69*, 497 (1975).
136. Hancock, W. S., Battersby, J. E.: Anal. Biochem. *71*, 260 (1976).
137. Hodges, R. S., Merrifield, R. B.: Anal. Biochem. *65*, 241 (1975).
138. Merrifield, R. B., Gisin, B. F., Bach, A. N.: J. Org. Chem. *42*, 1291 (1977).
139. Jakubke, H.-D., Klessen, Ch.: J. prakt. Chem. *319*, 159 (1977).
140. Hemmasi, B., Bayer, E.: Tetrahedron Lett. *19*, 1599 (1977).
141. Hagenmaier, H., Frank, H.: Hoppe-Seyler's Z. Physiol. Chem. *353*, 1973 (1972).
142. Birr, Chr., Flor, F., Fleckenstein, P., Lochinger, W., Wieland, Th.: In: Peptides 1971 (ed. H. Nesvadba), pp. 175–184. Amsterdam: North Holland Publ. 1973.
143. Wieland, Th., Birr, Chr., Flor, F.: Angew. Chem. *83*, 333 (1971); Angew. Chem. Int. Ed. Engl. *10*, 336 (1971).
144. Wieland, Th., Flor, F., Birr, Chr.: Liebigs Ann. Chem. *1973*, 1595–1600.
145. Flor, F., Birr, Chr., Wieland, Th.: Liebigs Ann. Chem. *1973*, 1601–1605.
146. Leuchs, H.: Ber. dtsch. Chem. Ges. *39*, 857 (1906).
147. Stelzel, P.: In: Houben-Weyl: Methoden der Organischen Chemie, Synthese von Peptiden (ed. E. Müller), Vol. 15/2, pp. 187–205. Stuttgart: Thieme 1974.
148. Miyoshi, M.: Bull. Chem. Soc. Jap. *43*, 3321 (1970); ibid. *46*, 212 and 1488 (1973).
149. Merrifield, R. B., Mitchell, A. R., Clark, J. E.: J. Org. Chem. *39*, 660 (1974).
150. Flor, F.: Doctoral Thesis, Heidelberg University, F. R. G. (1971).
151. Connert, W.-D.: Doctoral Thesis, Heidelberg University, F. R. G. (1976).
152. Bodanszky, M.: Acta Chim. Hungaria, *10*, 335 (1957).
153. Kovács, K., Penke, B.: In: Peptides 1972 (eds. H. Hanson, H.-D. Jakubke), pp. 187–188. Amsterdam: North Holland Publ. 1973.
154. Penke, B., submitted for publication.
155. Mukaiyama, T., Ueki, M., Maruyama, H., Matsueda, R.: J. Amer. Chem. Soc. *90*, 4490 (1968).
156. Maruyama, H., Matsueda, R., Kitazawa, E., Takahagi, H., Mukaiyama, T.: Bull. Chem. Soc. Jap. *49*, 2259 (1976).
157. Birr, Chr., Wengert-Müller, M., Buku, A.: In: Peptides (eds. M. Goodman, J. Meienhofer), pp. 510–513. New York: Wiley 1977.
158. Horiki, K.: Tetrahedron Lett. *22*, 1897 and 1901 (1977).
159. Matsueda, R., Maruyama, H., Kitazawa, E., Takahagi, H., Mukaiyama, T.: Bull. Chem. Soc. Jap. *46*, 3240 (1973).
160. Hagenmaier, H.-P.: Hoppe-Seyler's Z. Physiol. Chem. *356*, 777 (1975).
161. Birr, Chr., Frank, R.: FEBS Letters *55*, 68 (1975).
162. Tregear, G. W.: In: Peptides 1974 (ed. Y. Wolman), pp. 177–189. New York: Wiley 1975.
163. Van Rietschoten, J., Granier, C., Rochat, H., Lissitzky, S., Miranda, F.: Europ. J. Biochem. *56*, 35 (1975).
164. Birr, Chr., Frank, R.: FEBS Letters *55*, 61 (1975).
165. DeVries, J. X., Frank, R., Birr, Chr.: FEBS Letters *55*, 65 (1975).
166. Markley, L. D., Dorman, L. C.: Tetrahedron Lett. *1970*, 1787.

References

167. Wieland, Th., Birr, Chr., Wissenbach, H.: Angew. Chem. *81*, 782 (1969); Angew. Chem. Int. Ed. Engl. *8*, 764 (1969).
168. Penke, B., Birr, Chr.: Liebigs Ann. Chem. *1974*, 1999.
169. Mitchell, A. R., Merrifield, R. B.: J. Org. Chem. *41*, 2015 (1975).
170. Wang, S., Merrifield, R. B.: J. Amer. Chem. Soc. *91*, 6488 (1969).
171. Pietta, P. G., Cavallo, P. F., Marshall, G., Pace, M.: Gaz. Chim. Ital. *103*, 483 (1973).
172. Pless, J., Bauer, W.: Angew. Chem. *85*, 142 (1973); Angew. Chem. Int. Ed. Engl. *12*, 147 (1973).
173. Yajima, H., Ogawa, H., Watanabe, H., Fujii, N., Kurobe, M., Miyamoto, S.: Chem. Pharm. Bull. *23*, 371 (1975).
174. Matsuura, S., Nin, C. H., Cohen, J. S.: Chem. Commun. *1976*, 451.
175. Mitchell, A. R., Erickson, B. W., Ryabtsev, M. N., Hodges, R. S., Merrifield, R. B.: J. Amer. Chem. Soc. *98*, 7357 (1976).
176. Beyerman, H. C., Hindriks, H., deLeer, E. W. B.: Chem. Commun. *1968*, 1668.
177. Barton, M. A., Leumieux, R. U., Savoie, J. Y.: J. Amer. Chem. Soc. *95*, 4501 (1973).
178. Chang, J. K., Shimizu, M., Wang, S.: J. Org. Chem. *41*, 3255 (1976).
179. Wang, S.: J. Org. Chem. *41*, 3258 (1976).
180. Visser, S., Raap, J., Kerling, K. E. T., Havinga, E.: Rec. Trav. Chim. Pays-Bas *89*, 865 (1970).
181. Sieber, P., Kamber, B., Eister, K., Hartmann, A., Riniker, B., Rittel, W.: Helv. Chim. Acta *59*, 1489 (1976).
182. Shigezane, K., Mizoguchi, T.: Chem. Pharm. Bull. *21*, 972 (1973).
183. Glass, J. D., Meyers, Ch., Schwartz, J. L., Walter, R.: In: Peptides 1974 (ed. Y. Wolman), pp. 141–152. New York: Wiley 1975.
184. Wieland, Th., Birr, Chr., Fleckenstein, P.: Liebigs Ann. Chem. *756*, 14 (1972).
185. Veber, D. F., Milkowski, J. D., Denkewalter, R. G., Hirschmann, R.: Tetrahedron Lett. *1968*, 3057.
186. Weygand, F., Czendes, E.: Angew. Chem. *64*, 136 (1952).
187. Veber, D. F., Brady, St. F., Hirschmann, R.: In: Chemistry and Biology of Peptides (ed. J. Meienhofer), pp. 315. Ann Arbor U. S. A.: Ann Arbor Sci. Publ. 1972.
188. Birr, Chr., Fleckenstein, P., Wieland, Th.: Contribution to "Symposium on Peptide and Protein Chemistry, USSR-FRG"; Dushanbe, USSR, 1976; will be published elsewhere.
189. Rich, D. H., Gurwara, S. K.: J. Chem. Soc., Chem. Commun., *1973*, 610; J. Amer. Chem. Soc. *97*, 1575 (1975); Tetrahedron Lett. *1975*, 301.
190. Trietsch, P.: Doctoral Thesis, Heidelberg University, F. R. G. (1975).
191. Blecher, H., Pfaender, P.: Liebigs Ann. Chem. *1973*, 1263.
192. Brunfeldt, K., Halstrøm, J., Roepstorff, P.: In: Peptides 1968 (ed. E. Bricas), pp. 194–196. Amsterdam: North Holland Publ. 1968.
193. Loffet, A., Close, J.: In: Peptides 1968 (ed. E. Bricas), pp. 189–193. Amsterdam: North Holland Publ. 1968.
194. Voelter, W., Zech, K., Grubhofer, N.: Z. Naturforsch. *28b*, 625 (1973).
195. Bakalyar, S. R., Henry, R. A.: J. Chromatography *126*, 327 (1976).
196. Brunfeldt, K., Halstrøm, J., Roepstorff, P.: Acta Chem. Scand. *23*, 2830 (1969).
197. Mathur, N. K., Williams, R. E.: J. Macromol. Sci.-Rev. Macromol. Chem. *C15*, 117–142 (1976).
198. Gabriel, T. F., Jimenez, M. H., Felix, A. M., Michalewsky, J., Meienhofer, J.: Int. J. Peptide Protein Res. *9*, 129 (1977).
199. Wieland, Th., Rietzel, Chr.: Liebigs Ann. Chem. *754*, 107 (1971). Wieland, Th., Birr, Chr.: ibid. *757*, 136 (1972).
200. Colescott, R. L., Bossinger, C. D., Cook, P. J., Dailey, J. P., Enkoji, T., Flanigan, E., Geever, J. E., Groginsky, C. M., Kaiser, E., Laken, B., Mason, W. A., Olsen, D. B., Reynolds, H. C., Skibbe, M. O.: In: Peptides: Chemistry, Structure, Biology (eds. R. Walter, J. Meienhofer), pp. 463–467. Ann Arbor U. S. A.: Ann Arbor Sci. Publ. 1975.
201. Laursen, R. A.: In: Immobilized Enzymes, Antigens, Antibodies and Peptides (ed. H. H. Weetall), pp. 567–634. New York: M. Dekker 1975.
202. Wachter, E., Machleidt, W., Hofner, H., Otto, J.: FEBS Letters *37*, 217 (1973).
203. Cramer, F., Köster, H.: Angew. Chem. *80*, 488 (1968); Angew. Chem. Int. Ed. Engl. *7*, 473 (1968).
204. Köster, H.: Tetrahedron Lett. *16*, 1527–1538 (1972).

205. Atherton, E., Caviezel, M., Over, H., Sheppard, R. C.: J. Chem. Soc., Chem. Comm. *1977*, 819.
206. Gait, M. J., Sheppard, R. C.: Nucl. Acid Res. *4*, 1135 (1977).
207. Bodansky, M., Fagan, D. T.: Int. J. Peptide Protein Res. *10*, 375 (1977).
208. Reid, R. E.: J. Org. Chem. *41*, 1027 (1976).
209. Live, D. H., Agosta, W. C., Cowburn, D.: J. Org. Chem. *42*, 3556 (1977).
210. Kent, S. B. H., Mitchell, A. R., Barany, G., Merrifield R. B.: Analyt. Chem. *50*, 155 (1978).
211. Sato, K., Abe, H., Kato, T., Izumiya, N.: Bull. Chem. Soc. Jap. *50*, 1999 (1977).
212. Tregear, G. W., van Rietschoten, J., Sauer, R., Niall, H. D., Kentmann, H. T., Potts, Jr., J. T.: Biochem. *16*, 2817 (1977).
213. Sieber, P.: Helv. Chim Acta *60*, 2711 (1977).
214. Hruby, V. J., Upson, D. A., Agarwal, N. S.: J. Org. Chem. *42*, 3552 (1977).
215. Veber, D. F., Paleveda, Jr., W. J., Lee, Y. C., Hirschmann, R.: J. Org. Chem. *42*, 3286 (1977).
216. Schlatter, J. M., Mazur, R. H., Goodmonson, O.: Tetrah. Letters *33*, 2851 (1977).
217. Jones, Jr., D. A.: Tetrah. Letters *33*, 2853 (1977).
218. Bakkum, J. T. M., Beyerman, H. C., Hougerhout, P., Olieman, C., Voskamp, D.: Rec. Trav. Chim. Pay-Bas *96*, 301 (1977).
219. Burgus, R., Revier, J.: In: Peptides 1976 (ed. A. Loffet), pp. 85. Edit. Université de Bruxelles 1976.
220. Ciejek, E., Thorner, J., Geier, M.: Biochem. Biophys. Res. Comm. *78*, 952 (1977).
221. Gait, M. J., Sheppard, R. C.: Nucl. Acid Res. *4*, 4391 (1977).

Author Index

Author Index

Subject Index

Reactivity and Structure

Concepts in Organic Chemistry

Editors: K. Hafner,
J.-M. Lehn, C. W. Rees,
P. v. Ragué Schleyer,
B. M. Trost, R. Zahradnik

This series will not only deal with problems of the reactivity and structure of organic compounds but also consider synthetical-preparative aspects. Suggestions as to topics will always be welcome.

Springer-Verlag
Berlin
Heidelberg
New York

Volume 1: J. Tsuji

Organic Synthesis

by Means of Transition Metal Complexes
A Systematic Approach
1975. 4 tables. IX, 199 pages
ISBN 3-540-07227-6

Volume 2: K. Fukui

Theory of Orientation and Stereoselection

1975. 72 figures, 2 tables. VII, 134 pages
ISBN 3-540-07426-0

Volume 3: H. Kwart, K. King

d-Orbitals in the Chemistry of Silicon, Phosphorus and Sulfur

1977. 4 figures, 10 tables. VIII, 220 pages
ISBN 3-540-07953-X

Volume 4: W. P. Weber, G. W. Gokel

Phase Transfer Catalysis in Organic Synthesis

1977. 100 tables. XV, 280 pages
ISBN 3-540-08377-4

Volume 5: N. D. Epiotis

Theory of Organic Reactions

1978. 69 figures, 47 tables. XIV, 290 pages
ISBN 3-540-08551-3

Volume 6: M. L. Bender, M. Komiyama

Cyclodextrin Chemistry

1978. 14 figures, 37 tables. X, 96 pages
ISBN 3-540-08577-7

Volume 7: D. I. Davies, M. J. Parrott

Free Radicals in Organic Synthesis

1978. XII, 169 pages
ISBN 3-540-08723-0

Polymers

Properties
and
Applications

Editorial Board:
H.-J. Cantow,
H.J. Harwood,
J.P. Kennedy, J. Meißner,
S. Okamura, G. Olivé,
S. Olivé

Springer-Verlag
Berlin
Heidelberg
New York

Volume 1
B. Rånby, J.F. Rabek

ESR Spectroscopy in Polymer Research

1977. 356 figures, 29 tables. XIV, 410 pages
ISBN 3-540-08151-8

The main purpose of this book is to collect the present available information on the applications of electron spin resonance (ESR) spectroscopy in polymer research. The book has been written both for those who want an introduction to this field, and for those who are already familiar with ESR and are interested in application to polymers. Therefore, the fundamental principles of ESR spectroscopy are first outlined, the experimental methods including computer applications are described in more detail, and the main emphasis is on the application of ESR methods to polymer problems. The authors hope that this book will provide a useful source of information by giving a coherent treatment and extensive references to original papers, reviews, and discussions in monographs and books. In this way we hope to encourage polymer chemists, organic chemists, biochemists, physicists, and material scientists to apply ESR methods to their research problems. (2519 references).

Volume 2
H.-H. Kausch

Polymer Fracture

1978. Approx. 350 pages
ISBN 3-540-08786-9

In the last fifteen years modern spectroscopical methods (ESR, IR) and conventional methods of structure research have permitted considerable progress in the investigation of deformation and fracture of polymeric materials. For the first time in western languages a unified view of the kinetic theory of polymer fracture is presented by one of the scientists contributing to its development.

MIX
Papier aus verantwortungsvollen Quellen
Paper from responsible sources
FSC® C105338

If you have any concerns about our products,
you can contact us on
ProductSafety@springernature.com

In case Publisher is established outside the EU,
the EU authorized representative is:
**Springer Nature Customer Service Center GmbH
Europaplatz 3, 69115 Heidelberg, Germany**

Printed by Libri Plureos GmbH
in Hamburg, Germany